D1356605

Introduction
to Mathematical
Programming

Quantitative Tools
for Decision Making

Introduction to Mathematical Programming

Quantitative Tools for Decision Making

Benjamin Lev

Professor of Operations Research and
Chairman, Department of Management
Temple University
School of Business Administration
Philadelphia, Pennsylvania

Howard J. Weiss

Associate Professor of Operations Research,
Department of Management
Temple University
School of Business Administration
Philadelphia, Pennsylvania

Edward Arnold

First published in the United States of America in 1982
by Elsevier North Holland, Inc., New York 10017

First published in the United Kingdom in 1982
by Edward Arnold (Publishers) Limited, 41 Bedford Square, London WC1B 3DQ.

© 1982 by Elsevier North Holland, Inc.

ISBN 07131 3455 0

Desk Editor   Danielle Ponsolle
Designer   Glen Burris
Cover Design   Josê García
Production Manager   Joanne Jay
Compositor   Science Typographers, Inc.
Printer   Haddon Craftsmen

Manufactured in the United States of America

To our parents

Contents

Preface

Over the past three decades mathematical programming has become a widely used tool to aid managers with their decision making. Problems that thirty years ago were impossible to solve are now routine and solved by standard canned computer programs. The development of mathematical programming (and digital computers) has generated much interest in mathematical programming throughout businesses of all types and sizes. One recent indication of this interest is the inclusion of linear programming on actuarial examinations.

This book was written in order to introduce general business students and managers to the concepts, models, and methods of mathematical programming. The book is designed for the nonmathematician, with algebra being the only prerequisite. While the mathematical requirements are minimal, the reader will fully understand and appreciate the various methods of mathematical programming and the types of problems that can be solved by means of the mathematical programming algorithms.

For the novice who is entering the field, the book is designed to provide the conceptual framework necessary for understanding the terminology, models, and solution methods. Many detailed examples are included, some of which are carried over from chapter to chapter so that the reader will grasp the thoroughness of mathematical programming and appreciate the benefits and pitfalls of the discipline. Also, from the examples, the reader can readily see the application of the models to larger real-world problems. The algorithms are generally presented without rigorous mathematical proofs; instead, an outline of the proof is presented. The motivated reader who is accomplished in these areas can delve into the underlying mathematics by consulting the advanced references mentioned at the end of each chapter.

The book consists of ten chapters. Chapter 1 describes both the history of mathematical programming and the general characteristics of mathematical programming problems. Chapters 2 and 3 present the most widely used tool of mathematical programming—linear programming. In Chapter 2 a graphical discussion is presented, while in Chapter 3 an algebraic solution method for linear programming problems is presented. Chapter 4 discusses the immediate extensions that can be made when using linear programming. Chapters 5 and 6 are devoted, respectively, to two special cases of linear programming: the classical transportation problem and the assignment problem. Chapter 7 introduces integer programming problems and suggests three basic methods of solution. Chapter 8 presents a dynamic programming approach to problem solving and contains many numerical examples and potential applications of dynamic programming. Chapter 9 contains the more general model of nonlinear programming and emphasizes the search techniques that are used to solve these problems. Finally, Chapter 10 introduces the reader to advanced material on game theory, goal programming, and heuristic methods.

There are several ways that this text can be used in a course. Chapters 1 through 6 on linear programming should form the basis of any course. Additional material from Chapters 7 through 10 can be chosen according to the backgrounds of the students and the duration of the term. We have found that in a one-semester graduate (MBA) course the entire book can be covered (although barely), while in our undergraduate business course we omit Chapters 7 (integer programming) and 9 (nonlinear programming).

Since we have used this book at both the undergraduate and graduate levels, we are very grateful to the many students who suffered through and commented on earlier drafts of this text. In particular we would like to thank B. Orlando, Kathy Knoeble, and Edwina Smith for their many helpful suggestions. Also we would like to thank the Temple University Bureau for Economic and Business Research for typing this manuscript (several times) and the Management Department secretaries, Margaret Ciamaichelo and Debbie Czapka for their help with the preparation of this manuscript.

Finally, we would like to express our deep appreciation to our wives, Debbie and Lucia, and children, Ron, Nurit, Lisa, and Ernie for their understanding during the years that this manuscript was being prepared. We cannot atone for the many nights spent at Temple University rather than in front of the fireplace.

Introduction
to Mathematical
Programming
Quantitative Tools
for Decision Making

One

Introduction

1.1. A Historical Perspective

Since the conclusion of World War II, operations research—the application of scientific (especially mathematical) methods to problems involving complex systems (e.g., firm management or economic planning)—has been increasingly implemented as a tool for decision makers in government and private industry. Operations research in general and mathematical programming in particular facilitate decision making on policy issues that affect our daily lives. Airline and train scheduling and determining department store inventory levels are examples of areas in which operations research analyses benefit great numbers of people. Moreover, it is more important now than ever before that managers and decision makers fully understand operations research.

Unformalized operations research problems, including scientific approaches to organizational management problems, have been successfully solved for many decades. As a formalized structure, however, operations research was developed by the military during World War II.

The advent of the war spurred the efforts of the scientific community to contribute by collecting data and expanding its vision in order to come up with effective solutions to war-related problems. The need to allocate scarce resources so as to derive the maximum potential benefits compelled the military to acknowledge the scientific community's expertise in such matters. Hillier and Lieberman (1980, page 3) point out that the efforts of operations research teams of scientists were allegedly instrumental in winning the air battle of Britain, the island campaign in the Pacific, the battle of the North Atlantic, and so on.

It later became increasingly clear that the concepts utilized by the military could be extended to similar issues in a peacetime economy: Immediately following World War II, the proficiency of the military establishment captured the interest of managers who were searching for solutions to intricate industrial problems. At the same time, the scientific military community sought new areas to which to apply its knowledge and experience. The unprecedented postwar growth surge of the industrial sector was another factor contributing to the interest of business in techniques for coping with its problems. Both business and the military recognized that the problems confronting industry were similar in nature to those encountered during the war, so the concepts of operations research were put into an industrial perspective and were eventually extended to other areas as well.

In the professional literature operations research is generally defined as follows:

> Operations research is the application of scientific methods, techniques, and tools to problems involving the operations of a system in order to provide those in command with optimal solutions.

This approach consists of the following steps:

1. Understanding and describing the system;
2. Building a model of the real-life system;
3. Using the model as a basis for predicting future situations.

The last step includes the search for the optimal solution because in it the manager tries alternatives from which he chooses the optimum.

One of the principal areas and probably the most developed aspect of operations research is mathematical programming. This discipline includes linear programming, nonlinear programming, dynamic programming, integer programming, and other variants of programming problems. (Note that the term "programming" in these expressions has a different meaning from that intended in the phrase "computer programming.") *Mathematical programming* is a technique for determining the values of a set of decision variables that optimize a mathematical objective function and conform to a given set of mathematical constraints.

The first mathematical programming problem was formulated by Hitchcock (1941) and Koopmans (1951) in 1941 as a transportation or distribution problem. (A discussion of the transportation problem is presented in Section 1.3.3.) However, a good solution technique for the problem did not become available until 1947. In that year Dantzig (1963), while working for the United States Air Force, suggested an efficient way to solve this problem. A few years later, he generalized the transportation problem to what is now called linear programming. Since then, new techniques and applications have evolved through the efforts and cooperation of interested individuals in academic institutions and the industrial sector.

Nonlinear programming presents another perspective on mathematical programming problems. In 1951, Kuhn and Tucker (1956) introduced an innovative concept in nonlinear programming, thereby generating profound interest in the field among scientists. Within the succeeding decade an enormous collection of articles on this subject was published.

Dynamic programming was a concept that began to appear in 1957. The person who has contributed most to dynamic programming is Bellman, who published his first book on the subject in 1957.

Integer programming, which at first appeared to be a simple extension of linear programming, developed into a difficult concept. In 1958, Gomory (1963) suggested an acceptable approach to the solution of integer programming problems. Later, variations of this method appeared, but a practical mode for solving real-life, large-scale integer programming problems has not yet been devised.

Besides the growth of operations research theory, a second factor that contributed to the development of the field was the computer revolution. It became evident that providing effective solutions to the complex problems typically encountered in operations research requires an enormous amount of computation time—so much, in fact, that only a computer could meet the need. Hence, but for the extraordinary computational speed, storage capacity, and retrieval capabilities of modern digital computers, operations research would not have achieved its present status.

It has been noted that operations research provides support for decision making. It does so not only through the techniques previously mentioned, but also by borrowing from other sciences or, in some instances, by establishing a conceptual framework tailored to the specifications of the problem under investigation. Operations research offers the means for upper management to investigate possible consequences of the decision-making process. The general concepts of operations research have been useful in directing management toward a precise expression of its objectives, and demonstrate the way to achieve the desired results.

In recent years, both the number and the diversity of operations research applications have continued to increase rapidly. In fact, the impact of operations research appears unchallenged by that of any other recent development except the electronic computer. The extent of its effect is indicated by the number of academic institutions that offer operations research programs at various degree levels. In addition to this recognition by the academic community, operations research has acquired a successful performance record that has attracted the interest of many management consulting firms. The industries that have benefited from the concepts developed in this field are too numerous to mention. However, it is clear that the public and private sectors are just beginning to appreciate the importance of understanding and applying the concepts of operations research.

Mathematical programming is the most successful operations research technique. Linear programming has been applied successfully to problems relating to resource allocation, distribution, corporate planning, forecasting, health care delivery, and government strategy. Dynamic programming provides a different perspective on the search for solutions to problems involved with budgeting, scheduling, water supply systems, inventories, and control theory. Nonlinear programming has also been extensively employed in various areas when the objective and the restrictions are of a nonlinear form.

Decision making is a process present at all levels of society. Its effects may be beneficial or detrimental, depending on the awarenesses of those involved in the decision-making process. Operations research tends to make the task easier, using scientific, logical, objective studies to illustrate quantitative aspects and logical structure, so that the problem is clearly defined. Like many other applications of science, operations research is concerned with decision-making problems originating in real-life situations. For instance, most problems that confront local, state, and federal governments, economists, engineers, and industrialists concern the allocation of scarce resources. Operations research can provide considerable insight into the situations giving rise to such problems. However, we must be aware that it is not a crystal ball, and that its application requires considerable knowledge, experience, ingenuity, and innovation.

1.2. Structure of Mathematical Programming Problems

The goal of mathematical programming problems is to find values for the n decision variables $X=(x_1, x_2, \ldots, x_n)$. The decision maker must choose values for the variables so that he is in a position to optimize his objective which may include:

1. Maximize profit
2. Maximize utilization of equipment
3. Minimize cost
4. Minimize raw material or resources
5. Minimize traveling time

In most instances, there is only one objective to be optimized. However, there are cases where conflicting goals may exist in the same system. For example, we may want to maximize profit and at the same time maximize utilization of equipment. Therefore, an optimal solution to each of the two objectives may or may not be the same one. If the solutions are not the same, then we have to concentrate on one objective function and treat the other objective function as a constraint. For instance, we may want to maximize profit while maintaining a utilization level of at least 80% for the equipment. Let $f(X)$ denote the objective function to be optimized;

mathematically speaking, the goal in this case is to find X's to optimize $f(X)$ (opt. $f(X)$).

Frequently, we are not in a position to choose any variables X. In fact, there are some restrictions on the variables. Let us assume that the set of restrictions is S. The goal is to find a member of S, noted as $X \in S$, that optimizes $f(X)$. The structure of the set S and the function $f(X)$ determine the particular kind of problem to solve. S may have many shapes and various kinds of restrictions. The most common categories of problems are as follows:

1. Unconstrained optimization: Optimize $f(X)$ where the set S is the n-dimensional space. In other words, there is no restriction on the variable X. We are free to assign any values to the n decision variables $x_1, x_2, \ldots, x_n$.

2. Linear programming: $f(X)$ is a linear function $f(X) = c_1 x_1 + c_2 x_2 + \cdots + c_n x_n$ with constant coefficients c_j of the variables in the objective function. In this important category the set of restrictions S consists of only linear restrictions.

$$S = \left\{ X \mid \sum_{j=1}^{n} a_{ij} x_j \leq b_i, \quad i = 1, 2, \ldots, m \right\}$$

(This should be read as follows: The set S is the set of all X's that satisfy the linear constraints $a_{i1} x_1 + a_{i2} x_2 + \cdots + a_{in} x_n \leq b_i$ for constraint i, which ranges from 1 to m. (The a_{ij}, b_i are constant coefficients.)

3. Integer programming: This category requires that all decision variables assume only integer values: $x_j = 0, 1, 2, 3, 4, \ldots, j = 1, 2, \ldots, n$. Integer programming also requires the linear objective function and the linear set of constraints mentioned in 2. This statement is another restriction on S.

   A variant of this case is the 0–1 programming problem that obtains when $x_j = 0, 1, j = 1, 2, \ldots, n$, so that x_j can assume only the value 0 or 1. If some of the variables are continuous variables and some of the variables are integer variables, then the problem is called mixed integer programming.

Other restrictions on S can be of either a technological or a logical nature. For instance, a system may produce two items and the products might be associated with each other. In particular, a company could be restricted to produce sets of four chairs for every table. Another example might be a plant having the capacity to generate electricity and to distill salty water. A technological ratio may exist between the quantities of electricity and the water that it produces, for example, 1 kilowatt per 2 gallons of water. As another instance, it is impossible for a contractor to install a roof before the completion of the foundation and walls of a house. These examples represent technological restrictions that must be included in the set S.

In summary, the decision maker has to choose the variables $X = (x_1, x_2, \ldots, x_n)$. However, the choice is restricted to a specific set S. The set S contains all the restrictions that are specific to the problem. Among all possible solutions in S, the decision should be to choose the solution that optimizes the objective function $f(X)$. Mathematically, we can simply use the following notations:

$$\text{Optimize } f(X)$$
$$X \in S$$

1.3. Industrial Applications of Mathematical Programming

The best way to demonstrate the scope of mathematical programming is to show the wide variety of problems in the real world to which this model can be applied. Although hypothetical, these cases enable us readily to grasp the range and full dimensions of the problems that can be formulated and solved by mathematical programming.

The following examples have the common characteristics of one objective function and a list of constraints. These subjects are treated in detail in later chapters.

1.3.1. Resource Allocations

A furniture company in the South manufactures chairs and tables for distribution to retailers in various sections of the United States. The two resources necessary for producing the chairs or tables are wood (raw material) and man-hours. There is also a known specific profit for each chair and table that the company turns out. For the company to produce one chair or one table a certain amount of wood and man-hours is needed. In this case, the decision variables are the number of chairs and the number of tables that the company should produce. The objective function is to maximize the profit made by the chairs and tables, and the list of constraints should include the availability of the raw material and man-hours. Note that in this problem there is no ratio restriction of four chairs per table. However, this restriction could be easily included in the model if the decision maker considered the fact to be relevant.

1.3.2. Assignment Problem

A repair shop has 15 different machines, and each machine has the capability of performing various jobs. On a specific morning, the manager of the shop has to supervise the performance of 10 different jobs. Each assignment can be accomplished by any one of the 15 machines. The time it takes to perform job i on machine j is t_{ij}. It must be further noted that this time t_{ij} varies from machine to machine. In fact, if $t_{ij} = \infty$, then job i cannot be performed on machine j. In this case, the decision variables are the assignment of the jobs to the machines. The manager's objective is to determine an assignment of the 10 jobs to the 10 best machines so that the total time utilized by the machines is as small as possible. Furthermore, the constraint is that he must perform all 10 jobs in the shop. The manager may find himself in a restricted situation since he cannot simultaneously assign more than one job to a machine or more than one machine to a job.

1.3.3. Transportation Problem

The M. B. Miller beer company has six breweries in the eastern region of the United States. The beer produced at the six breweries shares the common characteristics of cost and quality. After the production process is completed, the beer is shipped to 100 major distributors located along the East Coast. Each brewery has a maximum capacity that cannot be exceeded and each distributor has a demand for beer that must be satisfied at a definite time. There is a transportation cost c_{ij} for shipping one gallon of beer from brewery i to distributor j. Therefore, the decision variables are x_{ij} for the amount of beer to be distributed from brewery i to distributor j. It is now evident that the objective function is to minimize total distribution costs, and the constraints are to meet distribution demand and plant capacity.

1.3.4. Integer Programming Problem

To illustrate this problem, let us consider a publisher who is most eager to promote the sale of a new book in the United States. Ten salesmen are available to distribute the book, and five regions have been designated for allocation purposes. An analysis of previous data indicates that each region is different from the others but tends to share a characteristic: An increased number of salesmen in any region will increase the amount of books that are sold there. However, there is a level of saturation after which the marginal contribution is relatively small. In this problem, the decision variables are the number of salesmen assigned to each region. Furthermore, the number can only be an integer, $x = 0, 1, 2, 3, 4, \ldots$. The objective function is to sell as many books as possible, and the constraints are the limited number of salesmen, the regions, and the behavior of the regions.

In addition, integer programming formulation is often employed in situations that have "on–off" characteristics. If the alternatives are to buy or not to buy, then the variable $x = 1$ is to buy, and $x = 0$ is not to buy. When the variables may have only 0 or 1 values, the problem is called a 0–1 programming problem. When some of the variables are integers and some are continuous, then it is called a mixed-integer programming problem.

1.3.5. Inventory Problem

A classic conflict in the inventory problem is between ordering large quantities or small quantities of an item. It is evident that a good reason for ordering large quantities is financial saving in the collective ordering of merchandise. Each order has a specific cost attached to the purchase, and reducing the number of orders (increasing the quantity of each order) yields a saving in total costs. At the same time, whenever we order large quantities, we invest a substantial amount of money in inventory. Therefore, the amount of capital available for other purposes is reduced. As a result, there is a tendency to maintain small inventories, thereby freeing more capital for other corporate expenditures.

The facts clearly indicate that a compromise will have to be reached in deciding whether to order large or small quantities. In this case, the variable is the number of items to be ordered at a specific time, and the objective function is to minimize total inventory cost for the stated policy. The restriction of this problem may be that the inventory level must always be nonnegative. Therefore, this inventory problem can be classified as a mathematical programming problem with constraints.

1.3.6. Dynamic Programming

An executive's managerial functions may include the allocation of funds for expansion. To be more specific, let us assume that a company has three plants and is considering the expansion of its overall capacity. The total budget for the three plants is listed as C and each plant can be expanded at three different levels. In addition, each expansion entails a projected revenue attached to the new facility. In this case, the decision variables are the levels of expansion of the plants. Furthermore, the objective function is to maximize the total revenue, and the budget C is the constraint. There are several approaches to solving this problem, but a very efficient one is the dynamic programming approach. Although the problem could be solved by explicitly enumerating all possible solutions, dynamic programming reduces the amount of computation.

1.4. Iterative Algorithms

Most of the solution methods in this book can be categorized as iterative algorithms. An algorithm is a procedure or set of sequential rules that can be used to arrive at a solution. "Iterative" means that the procedure is repeated several times. In the beginning, it is very important to provide a starting solution to the problem. A starting solution satisfies solution requirements but may not be optimal. At each iteration, the solution is revised, and after a finite number of iterations, the optimal solution is reached.

Consider, for example, the search for the tallest person among 100 given people. An iterative algorithm that can be used to solve this problem is the following:

1. Compare the first two persons and retain the taller of the two.
2. Take a person from among those not yet measured and compare this person with the one retained in Step 1.
3. Retain the taller of the two people compared in Step 2.
4. Terminate if all persons have been measured. Otherwise go to Step 2.

If there are more people to be compared, it is essential to continue to repeat Step 2. However, when no one remains to be compared, it is possible to terminate the procedure with the tallest person found in Step 2. Generally speaking, this simple iterative algorithm clearly demonstrates most of the properties of the iterative algorithms that will appear later in this book. The distinctive elements appearing in this algorithm are the following:

1. A starting solution is visible.
2. At each iteration there is an improvement and the number of iterations is finite.
3. A criterion exists for testing whether an optimal solution has been achieved.

These are the three important elements in any iterative algorithm, no matter how simple or complicated the problem. The following questions must be considered in judging any algorithm:

1. Will the starting solution be easy or difficult to find? In some instances the starting point is easy to find, but often finding a good starting solution may be as computationally difficult as the algorithm itself.
2. If there are several starting solutions, which is the best? The choice of starting solution affects the amount of work necessary for improving the starting solution.
3. How many iterations are needed to achieve optimality? This number plays a crucial role in choosing among algorithms. As the number of iterations increases, more calculations become necessary in order to achieve optimality. In addition, although it is usually guaranteed that an optimal solution will eventually be reached, the number of iterations may be infinite.
4. Occasionally the optimal solution is not required, and a solution near the optimum can be just as advantageous as an optimal one. If it is possible to indicate that the most recent solution is within 5% of the optimal solution, the answer may then be satisfactory and a further search is not necessary. In addition to the current solution, some algorithms also provide the gap between the solution on hand and the optimal solution.

5. The last important factor to be considered is the availability of computer programs and the existing facilities needed to solve the problem. Even though learning algorithms can be tedious, cumbersome, and exhausting, once you understand the procedure itself, you will be in a position to solve the problem with the aid of a computer. In choosing an algorithm to solve a problem, it is important to consider its capability to solve a problem of the size with which you are dealing, the storage capacity of your computer, and the time needed to solve the problem.

1.5. Summary and Plan of the Book

The growth of both mathematical programming and computer science over the last thirty years has made decision making very straightforward in several areas. Many problems are now standard and are simply solved using one or more of the many software packages that are available for mathematical programming. Examples of some of these problems were presented in Section 1.3, and the remaining chapters of this book present more detailed analyses of these problems and, of course, the solution methods.

In Chapters 2, 3, and 4 linear programing is examined in detail. Several examples are given of problems that are amenable to solution by means of linear programming. In chapter 2 a geometrical interpretation of linear programming is presented, while in Chapter 3 the algebraic solution method is given. In Chapter 4, extensions to linear programming are examined.

The transportation and assignment examples of Section 1.3 are special types of linear programming problems. As such, there are more efficient ways of solving them than the standard algebraic methods. The distinctive features of the two problems and the appropriate solution methods are presented in Chapters 5 and 6.

When dealing with problems that are restricted to having integer solutions, special methods must be used. In Chapter 7 three integer programming methods are presented and in Chapter 8 the dynamic programming approach is offered. Similarly, if a problem contains nonlinear functions, then nonlinear methods must be used. Some of these approaches are presented in Chapter 9.

Finally, in Chapter 10 three topics of general interest are presented. These topics should entice you to continue studying mathematical programming.

The mathematics used for the analyses in this book has been kept simple in order to facilitate your understanding and appreciation of the types of problems and the solution methods presented. The only mathematical requirements are the basic operations of addition, subtraction, multiplication, and division. A more sophisticated presentation can be found in the references listed at the end of each chapter. (An excellent general reference for operations research is Hillier and Lieberman, 1980.)

References and Selected Readings

Ackoff, R. L., and M. W. Sasieni 1968. *Fundamentals of Operations Research*. New York: John Wiley and Sons, Inc.

Bellman, R. F. 1957. *Dynamic Programming*. Princeton, New Jersey: Princeton University Press.

Dantzig, G. B. 1963. *Linear Programming and Extensions*. Princeton, New Jersey: Princeton University Press.

Gomory, R. E. 1963. An algorithm for integer solutions to linear programs. In *Recent Advances in Mathematical Programming*, ed. Robert L. Graves and Philip Wolfe, New York: McGraw-Hill.

Hillier, F. S., and G. J. Lieberman 1980. *Introduction to Operations Research*, 3rd ed. San Francisco, California: Holden-Day, Inc.

Hitchcock, F. L. 1941. The distribution of a product from several sources to numerous localities. *J. Math. Phys.* 20:224–230.

Koopmans, T. C. 1951. *Activity Analysis of Production and Allocation*. New York: John Wiley and Sons, Inc.

Kuhn, H. W., and A. W. Tucker 1956. *Linear Inequalities and Related Systems*. Princeton, New Jersey: Princeton University Press.

Wagner, H. M. 1975. *Principles of Management Science*, 2nd ed. Englewood Cliffs, New Jersey: Prentice-Hall, Inc.

Two | Linear Programming— The Graphical Solution

2.1. Introduction

Linear programming is the most important category of mathematical programming. Since World War II, the development of linear programming has been tremendous. It is evident that this theory has affected many of the executive's daily decision-making processes. Some compare the impact of linear programming on modern scientists to that of calculus on 17th-century scientists. Today there are very few quantitative books in any area of business or science that do not include the term linear programming in their indexes. There are two basic reasons for the spread of linear programming. The first is the wide variety of problems that can be adapted to and solved by the technique. Problems arise in areas such as finance, accounting, and chemistry, and in various phases of decision making, such as planning, design, and control. The need for linear programming exists at all levels of management as well as in research and industry.

Whereas linear programming problems are easily solved, nonlinear programming problems are relatively difficult to handle. Therefore, many nonlinear programming problems are forced into linear programming form, either by changing the assumptions or by replacing a nonlinear curve with a piecewise linear one, and are then solved by linear programming.

The second reason for the success of linear programming is its efficiency as a problem-solving technique. The operations research literature is saturated with articles on the theory of linear programming. Numerous algorithms, some of a general nature and others that are very specific, having been designed for special cases, are available for application to linear programming problems. Efficient commercial computer programs, which save both time and money for the consumer, are also available. The existence of a relatively easy procedure by which to obtain a solution prompts the decision maker to implement the technique as often as possible.

2.2. Basic Properties of Linear Programming

Linear programming is a method of determining the optimal values of a set of variables when linear combinations of the variables must lie within certain bounds. By *optimal values* we mean those that either maximize or minimize (depending on the type of problem) the value of a given linear relationship of the variables, called the *objective function*. The term objective function is applied because this is the target we are trying to reach. For example, the objective function is an equation that can represent either profit (to be maximized) or cost (to be minimized). The bounds referred to in this definition are represented by a set of functions called *constraints*. The general mathematical programming problem is to find the variables

$$X = (x_1, x_2, \ldots, x_n)$$

that optimize the objective function $f(X)$ subject to a set $g_i(X)$ of m constraints:

$$g_i(X) \left(\begin{matrix} \leq \\ = \\ \geq \end{matrix} \right) b_i \qquad i = 1, 2, \ldots, m$$

The symbol $\left(\begin{matrix} \leq \\ = \\ \geq \end{matrix} \right)$ means that $g_i(X)$ can be either less than or equal to ($\leq$), equal to ($=$), or greater than or equal to ($\geq$) b_i. The mathematical presentation is as follows.

> Find $x_1, x_2, \ldots, x_n$ (a solution)
> that optimize $f(X)$ (the objective function)   (2.1)
> subject to $g_i(X) \left(\begin{matrix} \leq \\ = \\ \geq \end{matrix} \right) b_i$ $i = 1, 2, \ldots, m$ (set of constraints)

In the case of linear programming, both the objective function and the constraints must be linear functions. (Linear functions are explained later in this section.) For instance, let the objective function be $Z = f(X)$ where

$$Z = c_1 x_1 + c_2 x_2 + \cdots + c_n x_n$$

and the set of constraints is

$$a_{i1} x_1 + a_{i2} x_2 + \cdots + a_{in} x_n \left(\begin{matrix} \leq \\ = \\ \geq \end{matrix} \right) b_i \qquad i = 1, 2, \ldots, m$$

Furthermore, let us use the symbol Σ for summation of a sequence of elements: For example,

$$\sum_{i=1}^{10} k_i = k_1 + k_2 + \cdots + k_{10}$$

The linear programming problem is now written as follows.

Find $\qquad\qquad x_1, x_2, \ldots, x_n$

$$\text{that optimize} \qquad Z = \sum_{j=1}^{n} c_j x_j \qquad\qquad (2.2)$$

$$\text{subject to} \qquad \sum_{j=1}^{n} a_{ij} x_j \left(\begin{smallmatrix} \leq \\ = \\ \geq \end{smallmatrix} \right) b_i \qquad i = 1, 2, \ldots, m$$

To be more specific, Problem (2.2) has n decision variables x_j, a linear objective function, and a set of m linear constraints. It can also be noted that all coefficients c_j, a_{ij}, and b_i are known constants.

All linear functions have two important properties. In fact, any function that has the following two properties may be called a linear function.

Property 1. Divisibility. In reviewing the facts, notice that the objective function of Problem (2.2) is stated as $Z = \sum_{j=1}^{n} c_j x_j$, and that the contribution of variable x_j to the objective function is proportional to c_j. If $x_j = 1$, the contribution is c_j. If $x_j = 2$, the contribution is $2c_j$. The fact that we can choose any value for x_j does not affect the proportion c_j in its contribution to the objective function. In linear programming problems, the discount of large quantities is nonexistent (i.e., there is no reduction in unit costs as production increases, since this would make the relationship nonlinear).

Property 2. Additivity. The linear function $Z = \sum_{j=1}^{n} c_j x_j$ has the property that if we choose $x_1 = 1$ and $x_2 = 1$, the total contribution of the two variables is simply $c_1 + c_2$. There is no discount because we order both $x_1 = 1$ and $x_2 = 1$.

These two properties are explained once more in the context of the next example.

The Exclusive Furniture Company earns a profit of $5 per chair and $7 per table. If the company makes x_1 chairs and x_2 tables, the total profit or contribution is $5x_1 + 7x_2$. The raw material needed to complete one chair is 2 board feet (B.F.) and the required raw material necessary to finish one table is 5 B.F. In addition, the man-hours required for production are 3 for each chair and 2 for each table.

The two properties can now be explained as follows: Property 1 means that if contribution to profit for one chair is $5, then for two chairs it is $2 \times 5 = \$10$ and for 100 chairs it is $500 (without any increase because of economies resulting from more productivity of larger quantities or without any discount). The amount of wood needed for one chair is 2 B.F.; for six chairs it is $6 \times 2 = 12$ B.F. (without discount). The amount of time needed to make one chair is 3 man-hours; to complete four chairs, $4 \times 3 = 12$ man-hours would be required (without discount). This same property holds true for the table.

Property 2 indicates that if we make 10 chairs and 6 tables, the total contribution to profit is $10 \times 5 + 6 \times 7 = \$92$. The raw material needed is $10 \times 2 + 6 \times 5 = 50$ B.F., and the time necessary for completion of the task is $10 \times 3 + 6 \times 2 = 42$ man-hours.

It should be noted that in linear programming the data are clearly stated. The coefficients a_{ij}, b_i, c_j are given and are known parameters that cannot be changed. This is an important assumption because it means that the model is deterministic or predictable. There is complete certainty with respect to the outcome of any process. For example, it may be likely that inflation will affect the price of a chair. However, the $5 profit is a fixed and given profit, perhaps suggested by a marketing department, and does not vary. The assumption enables the manager to solve the problem and arrive at an optimal solution. The model is of a static nature and does not permit any dynamic changes in the data. Other models have the capability for changes of input data and for complete parameterization of all data. This topic will be covered in Section 4.5.

2.3. Graphical Solution—A Maximization Problem

Let us continue with the problem presented in Section 2.2 but change the decision variables from x_1, x_2 to x, y, respectively. As previously stated, the Exclusive Furniture Company produces x number of chairs and y number of tables. In this case the objective of the firm is to maximize the profit $P = 5x + 7y$. However, there are two restrictions that must be applied to the production process. One restriction is on the raw materials and the other is on the man-hours needed to produce the chairs and tables. Furthermore, let us assume that for the coming week the restriction on board feet is 50 B.F., and the amount of man-hours available is 42 man-hours. Then the first restriction—that on the raw material—can be written as

$$2x + 5y \leqslant 50$$

(Recall that 2 B.F. are needed for each chair and 5 B.F. for each table.) The left-hand side clearly indicates the actual amount of raw material that is going to be used during the next week, and the right-hand side indicates the limit that this amount cannot exceed.

The second constraint is the man-hours restriction, which can be expressed as

$$3x + 2y \leqslant 42$$

(Recall that 3 hours are needed for each chair, 2 for each table.) The left-hand side is the number of man-hours that are going to be utilized in the production process and the right-hand side indicates the maximum available man-hours for the week.

Since x and y are the number of chairs and tables, respectively, to be produced, it is evident that $x \geq 0$ and $y \geq 0$. (If x were less than 0, it would indicate that we would produce a negative amount of chairs, i.e., would disassemble existing chairs for the raw material.)

Thus far, the Exclusive Furniture Company problem has been presented in a descriptive form, but now the problem can be translated into mathematical language. This process is called *mathematical modeling*. Without any loss of the details involved in this case, we can summarize the problem as follows.

Find x and y that maximize $P = 5x + 7y$ (objective function) while satisfying the restrictions:

$$2x + 5y \leq 50 \qquad \text{(material)}$$

$$3x + 2y \leq 42 \qquad \text{(man-hours)} \qquad\qquad (2.3)$$

$$x \geq 0, \ y \geq 0 \qquad \text{(nonnegativity)}$$

It becomes clear that the solution x, y to Problem (2.3) is the solution to the Exclusive Furniture Company given in its descriptive form.

The important fact to remember is that the first step in the solution of any mathematical programming problem is *the conversion of the problem from a descriptive form to a mathematical form*. Then, the goal is to solve the set of linear constraints and the linear objective function.

The activity space for Problem (2.3) is all possible values that can be associated with x and y. If we plot the two axes x and y, as indicated in Figure 2.1, then the two-dimensional space—or in this case the plane—contains all possible values for x and y.

Figure 2.1. Two-Dimensional Activity Space

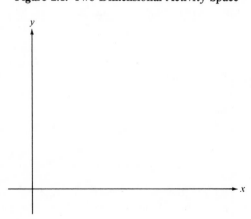

From the previous discussion, it is evident that not all the points in the x-y plane satisfy the constraints of Problem (2.3). For example, the point $x=1000$ and $y=5000$ violates the constraints because the raw material and man-hours given are not sufficient to produce 1000 chairs and 5000 tables. The next step is to define clearly the region that does satisfy the constraints.

Raw material restriction. The previously stated raw material restriction for this problem is $2x+5y\leqslant50$. Instead of working with an inequality, we can simplify the situation and use the equation $2x+5y=50$ to arrive at a workable solution. The line intersects the x and y axes at the following points:

For $x=0$	For $y=0$
$2\cdot0+5\cdot y=50$	$2\cdot x+5\cdot0=50$
$5y=50$	$2x=50$
$y=10$	$x=25$

Thus, the two intersection points are $(0,10)$ and $(25,0)$.

In Figure 2.2 the two intersection points are plotted as well as the straight line that passes through them. It is evident that this line is the line $2x+5y=50$. In fact, since two points determine a straight line, any point on this graph satisfies the equation $2x+5y=50$, and any point satisfying the equation also is on the line. For instance, the point $(x=5, y=8)$ satisfies the equation $2\cdot5+5\cdot8=50$, and this point is also on the line. The significant

Figure 2.2. Exclusive Furniture Company Problem—Raw Material Restriction

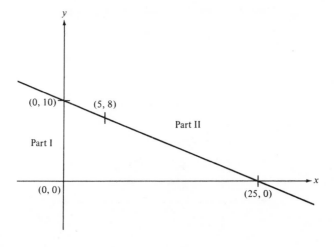

concept is that the equation is an algebraic representation of the designated line while the graph is a geometric representation of the same phenomenon. Therefore, the two are equivalent. In addition, note that the line in Figure 2.2 divides the two-dimensional space into two parts, Part I and Part II, called half planes. Furthermore, recall that the original constraint is the inequality $2x+5y\leqslant50$, and that any point in Part I satisfies the inequality.

The shaded region in Figure 2.3 is the one that satisfies the constraint. For verification check the following points: The origin

$$P_1=(0,0)$$

$$2\cdot0+5\cdot0\leqslant50$$

or the point $x=2$, $y=-3$

$$P_2=(2,-3)$$

$$2\cdot2+5\cdot(-3)\leqslant50$$

$$-11\leqslant50$$

both of which satisfy the constraint. You can verify that other points in the half plane labeled Part I also satisfy the restriction $2x+5y\leqslant50$. At this time, note that any point in Part II violates the stated constraint. For example, $x=20$, $y=15$

$$P_3=(20,15)$$

$$2\cdot20+5\cdot15=115>50$$

The points on the line $2x+5y=50$ also satisfy the constraint. To summarize the analysis, an inequality constraint divides the plane into two parts. To

Figure 2.3. Exclusive Furniture Company Problem—Raw Material Acceptable Region

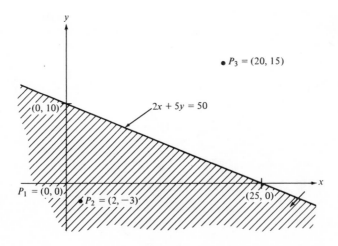

decide which part contains points that satisfy the constraint, simply choose a point in one of the parts. If the chosen point satisfies the constraint, then the entire half plane in which the point is located, including the dividing line, is the acceptable region. This is the shaded region in Figure 2.3. Consequently, the complementary region is the set of all points that violate the constraint. Generally speaking, this specific area is usually signified by an arrow, as shown. The direction of the arrow designates the half plane in which all the points satisfy the constraint.

Man-hours restriction. The man-hours restriction $3x+2y \leqslant 42$ can be analyzed in a similar way. Instead of an inequality, apply the equality $3x+2y=42$ to determine the dividing line of the plane. The line $3x+2y=42$ intersects the axes at the following points:

For $x=0$	For $y=0$
$3 \cdot 0 + 2 \cdot y = 42$	$3 \cdot x + 2 \cdot 0 = 42$
$2y = 42$	$3x = 42$
$y = 21$	$x = 14$

The line $3x+2y=42$ passes through the two points $(0,21)$ and $(14,0)$ and divides the plane into two regions, Part I and Part II. As before, it is sufficient to test only one of the points to determine which part satisfies the constraint. If we arbitrarily pick the origin $(0,0)$, and then substitute the coordinates into the inequality, $0+0 \leqslant 42$, the results indicate that the origin satisfies the constraint. Thus, Part I, including the dividing line, satisfies the man-hours restriction, and any point in Part II therefore violates the constraint. This fact is indicated in Figure 2.4 by both the shading and the arrow.

Superimposing the two constraints of Figures 2.3 and 2.4 on one illustration results in Figure 2.5. A brief examination of this figure reveals that the two restrictions have generated four distinct regions, labeled I, II, III, and IV. Furthermore, notice that the four regions violate or satisfy the constraints; this is summarized in the following table:

	Constraints	
Any point in	Raw material	Man-hours
I	Violates	Violates
II	Violates	Satisfies
III	Satisfies	Satisfies
IV	Satisfies	Violates

Since it is required to satisfy all the restrictions at the same time, the next step is to concentrate on the intersection of the regions in which all the constraints are satisfied. In this case, the facts indicate that region III

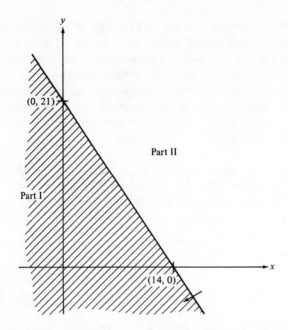

Figure 2.4. Exclusive Furniture Company Problem—Man-hours Acceptable Region

Figure 2.5. Exclusive Furniture Company Problem—Two-Constraint Acceptable Region

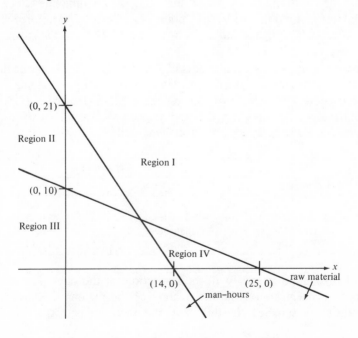

satisfies both restrictions. Therefore, it is safe to assume that the optimal solution must be located in this area.

The remaining two constraints, $x \geqslant 0$ and $y \geqslant 0$, limit the search to the first quadrant.

All four restrictions are summarized in Figure 2.6. It is essential to note that we have now designated another point, the intersection point of the equations

$$2x + 5y = 50$$
$$3x + 2y = 42$$

Multiply the first equation by 3 and the second equation by 2 and subtract the second equation from the first.

$$6x + 15y = 150$$
$$\underline{6x + 4y = 84}$$
$$11y = 66$$
$$y = 6$$

Now substitute $y = 6$ into the original first equation.

$$2x + 5 \cdot 6 = 50$$
$$2x = 20$$
$$x = 10$$

Figure 2.6. Exclusive Furniture Company Problem—Feasible Region

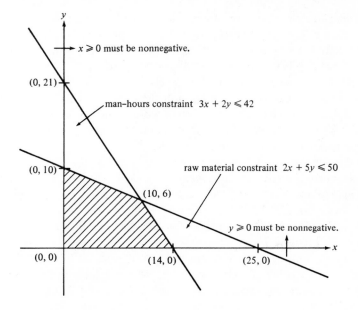

The intersection point of the above two equations is $x=10$, $y=6$.

Before we proceed, we define a few of the concepts presented so far and illustrate the definitions with examples.

Intersection Point. A point at an intersection of any two constraints is an intersection point. For instance, the point $(0,21)$ is the intersection of $3x+2y=42$ and the line $x=0$. Other intersection points are $(0,10)$, $(10,6)$, $(0,0)$, $(14,0)$, and $(25,0)$. There are six intersection points because there are four constraints and the number of ways to choose two different constraints at a time is the combination of four items taken two at a time, designated as

$$\binom{4}{2} = \frac{4!}{2!(4-2)!} = 6$$

Feasible Region. The shaded area that contains all the points that meet all the constraints is called the feasible region. This region is the most important area in solving the problem. As a matter of fact, one of the points in the feasible region is the optimal solution; outside the feasible region, there is no point that satisfies the restrictions.

Extreme Point. The set of intersection points that are also feasible points constitutes the set of extreme points. (In Figure 2.6 these are $(0,10)$, $(10,6)$, $(0,0)$, and $(14,0)$).

So far our discussion has focused on the feasible region. Next, we discuss the objective function.

The objective function. For our problem, the objection function is $f(x, y)$ $=$ maximize $p=5x+7y$. Let us assume that the projected profit for this line is \$70. There are several combinations of chairs and tables that contribute a profit of $p=$\$70. For example, $x=14$, $y=0$ contributes $5\cdot14+7\cdot0=$\$70. In other instances, other combinations, such as $x=0$, $y=10$ or $x=7$, $y=5$, contribute \$70. Observe that on a graph these three points, as indicated in Figure 2.7, are on the same line. In other words, this line is the line with the equation $5x+7y=70$. Any point on this line has the property of contributing exactly \$70 profit for the production line.

For the sake of discussion, let us now assume that the desired contribution has been changed by management to \$140. Some combinations of chairs and tables that yield \$140 are $x=28$, $y=0$; $x=0$, $y=20$; and $x=14$, $y=10$. Once again, the three points are on a line with the equation $5x+7y=140$. This isoprofit line (the line at which the profit is the same) is indicated in Figure 2.7.

It is interesting to note that the two profit lines of \$70 and \$140 are parallel to each other. In fact, the slope of these lines is determined by the coefficients of the variables, which are 5 for x and 7 for y and which

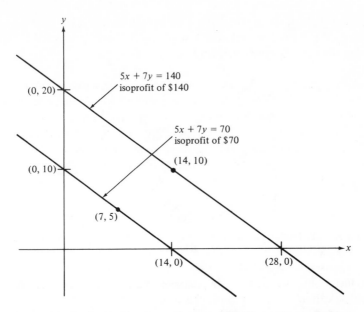

Figure 2.7. Exclusive Furniture Company Problem—Isoprofit Lines

remain the same for all the isoprofit lines. The most important point to note is that *all* isoprofit lines are parallel to each other. The only difference among the isoprofit lines is that the farther an isoprofit line is from the origin, the larger the profit contribution for that line is, and this is true whenever the coefficients of x and y are positive. As a result, since the object is to achieve the highest possible profit, it is desirable to move the isoprofit line as far from the origin as possible.

A clear way to take advantage of this property can be seen in Figure 2.8, where an isoprofit line is drawn in the feasible region. We want to determine the direction in which the objective function is optimized. Whenever the coefficients of the variables are positive, the farther the isoprofit line is from the origin, the higher the profit is. Recall that the arrow on the objective function indicates the desired direction. Now move the isoprofit line parallel to itself in the desired direction and do not stop until it leaves the feasible region. (You can do this by placing a pencil on the isoprofit line $Z=35$ and rolling it toward the upper right-hand corner of the page.) *Finally, it is important to understand that the last point in the feasible region touched by a line parallel to the original isoprofit line is the optimal solution.* Furthermore, this line always touches an extreme point last. The reason for this is that it is the feasible solution with the highest profit. Figure 2.8 indicates that the point $x=10$, $y=6$, with the profit

$$5x+7y=5\cdot 10+7\cdot 6=\$92$$

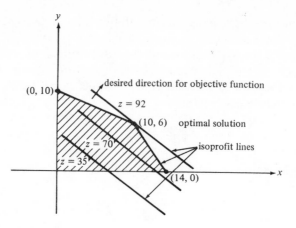

Figure 2.8. Exclusive Furniture Company Problem—Optimal Solution

is the last point touched in the feasible region. Therefore, $(10, 6)$ is the optimal solution.

2.4. A Minimization Problem

Many business and industrial problems are grouped in a family of problems known as "diet problems." (Despite this designation, the problems in this group are not necessarily related to food, vitamins, or nutrition.) Diet problems are concerned with any entities that can be selected in various quantities by choosing, combining, or mixing the elements. As an example we consider a diet problem that does relate to the specific issue of diet: A vegetarian diet may limit an individual to the selection of one vegetable and one juice. Once a week, the dieter purchases x units of vegetable and y units of juice. The shopper is further constrained in choosing this food by a requirement to consume a specific number of grams of vitamins. In this case, the purchases must include 40 grams of vitamin A, 50 grams of vitamin B, 70 grams of vitamin C, 10 grams of vitamin D, and 60 grams of vitamin E. In each unit of vegetable there are 0.1, 0.2, 0.04, 0.1, and 0.06 of a gram of vitamins A, B, C, D, and E, respectively. In each unit of juice, however, there are 0.05, 0.15, 0.2, 0.1, and 0.1 of a gram of vitamins A, B, C, D, and E, respectively. The price of one unit of vegetable is 2 cents and the price of one unit of juice is 3 cents. The vegetarian's problem, then, is to purchase quantities of the vegetable and juice such that his vitamin requirements are satisfied and the cost is as low as possible. In this instance, if the vegetarian is compelled to purchase x units of vegetable and y units of juice, then the price in dollars that the dieter is expected to pay is $0.02x + 0.03y$ (because a unit of vegetable and a unit of juice have specific costs of 2 and 3 cents, respectively). In addition, the list of constraints consists of meeting the requirements for all five vitamins, as stated earlier in the problem. Since each unit of vegetable contains 0.1 gram of vitamin A, by purchasing x units the individual buys $0.1x$ grams of vitamin A. In addition, each unit of juice contains 0.05 gram of vitamin A, and by purchasing y units, the dieter obtains $0.05y$ grams of vitamin A. Therefore, in the total purchase of vegetable and juice, the person has the ability to buy $0.1x+0.05y$. This quantity must be at least 40 grams, which is the requirement for vitamin A. Thus the constraint for vitamin A is expressed as $0.1x+0.05y \geqslant 40$.

In a similar way, the restrictions for vitamins B, C, D, and E can be written as follows.

vitamin B	$0.2x+0.15y \geqslant 50$
vitamin C	$0.04x+0.2y \geqslant 70$
vitamin D	$0.1x+0.1y \geqslant 10$
vitamin E	$0.06x+0.1y \geqslant 60$

Since x and y are the quantities bought, then obviously $x \geqslant 0$, $y \geqslant 0$.

The mathematical problem of the vegetarian dieter can now be summarized:

Find x and y that minimize
$$c=0.02x+0.03y$$
subject to
$$0.1x+0.05y \geqslant 40$$
$$0.2x+0.15y \geqslant 50$$
$$0.04x+0.2y \geqslant 70 \qquad (2.4)$$
$$0.1x+0.1y \geqslant 10$$
$$0.06x+0.1y \geqslant 60$$
$$x \geqslant 0 \; y \geqslant 0$$

The solution method for this problem is analogous to the solution method in Section 2.3. The restrictions are plotted in Figure 2.9. In order to find the intersection points of the first constraint (vitamin A) with the axis, let us begin with $x=0$ and $y=0$.

Converting the constraint from an inequality to an equation results in $0.1x+0.05y=40$. Then, solving for $x=0$, we substitute 0 for x in the stated equation:

$$0.1 \cdot 0+0.05 \cdot y=40$$
$$0.05y=40$$
$$y=800$$

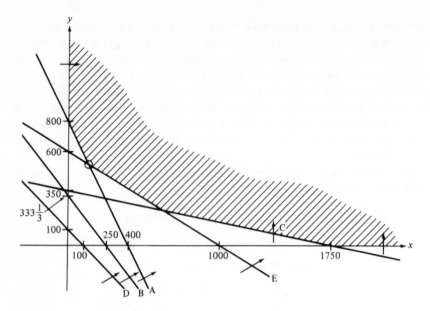

Figure 2.9. The Diet Problem—Feasible Region

For $y=0$, the equation is given as

$$0.1 \cdot x + 0.05 \cdot 0 = 40$$
$$0.1x = 40$$
$$x = 400$$

The results of the intersection points of this method are summarized in the following table:

Constraint	Intersection with the y axis $x=0$		Intersection with the x axis $y=0$		Coordinates
A $0.1x+0.05y=40$	$0.05y=40$	$y=800$	$0.1x=40$	$x=400$	$(0,800);(400,0)$
B $0.2x+0.15y=50$	$0.15y=50$	$y=333\frac{1}{3}$	$0.2x=50$	$x=250$	$(0,333\frac{1}{3});(250,0)$
C $0.04x+0.2y=70$	$0.2y=70$	$y=350$	$0.04x=70$	$x=1750$	$(0,350);(1750,0)$
D $0.1x+0.1y=10$	$0.1y=10$	$y=100$	$0.1x=10$	$x=100$	$(0,100);(100,0)$
E $0.06x+0.1y=60$	$0.1y=60$	$y=600$	$0.06x=60$	$x=1000$	$(0,600);(1000,0)$

Recall that the original constraints are of an inequality nature, and in order to determine which part of the plane is met by the constraints it is sufficient to check only one point. For example, let us begin with the origin $(0,0)$. After analyzing the original constraints, we see that the origin violates the five restrictions. The origin means purchase $x=0$ units of vegetable and $y=0$ units of juice, and with these amounts each of the five restrictions on vitamins is violated. The arrow on each constraint in Figure 2.9 indicates the direction in which any point on that half plane satisfies the specific constraint. Since the consumer expects to satisfy all five constraints simultaneously, it is the area of intersection of the permitted half plane that determines the final region. In addition, consider that $x \geqslant 0$ and $y \geqslant 0$. Consequently, the region that meets the total number of constraints is the shaded region in Figure 2.9.

Interestingly, the feasible region in this problem is of an unbounded nature, in the sense that the feasible region is open ended in one direction. Before continuing to determine the optimal solution, consider two definitions.

Unbounded feasible region. An unbounded feasible region is not surrounded by restrictions, and in at least one direction, one variable may be as large as desired and still satisfy all restrictions. For example, in the diet problem, $x = 1,000,000$, $y = 50$ is a feasible solution.

The restriction of vitamin D appears to be unimportant. Overlooking this restriction does not affect the feasible region. In fact, by the same argument, the constraint for vitamin B is also unimportant. However, any movement of the constraints for vitamins A, C, or E, and $x \geqslant 0$ and $y \geqslant 0$, changes the feasible region. Thus the following definitions are made:

Redundant Constraint. A redundant constraint is one that has no effect on the shape of the feasible region. Any nonredundant constraint defines a part of the feasible region boundary.

Active Constraint. A constraint is active with respect to a specific feasible point if the feasible point lies on the boundary of the constraint.

Determining the optimal solution. Let us now continue with the problem, adding the assumption that the vegetarian can spend \$6 on food for the following week. Two combinations of a vegetable and a juice that comply with this budget are

$$x = 300 \qquad y = 0 \qquad \text{cost } \$6$$

and

$$x = 0 \qquad y = 200 \qquad \text{cost } \$6$$

In fact, any combination of x and y that satisfies $0.02x + 0.03y = 6$ costs exactly \$6. The isocost line (the line for which the cost is constant) of \$6 is indicated in Figure 2.10. Unfortunately, none of the points on the isocost line of \$6 intersect the feasible region. This fact means that it is impossible to satisfy all the requirements within the budget of \$6.

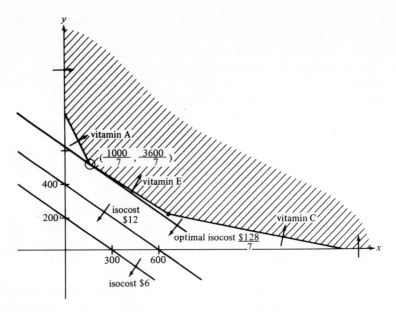

Figure 2.10. The Diet Problem—Isocosts Lines and Optimal Solution

Therefore the next attempt proceeds to the level of $12. The isocost line for this equation is expressed as

$$0.02x + 0.03y = 12$$

It can be shown that this line intersects the axes at $x = (12/0.02) = 600$ and at $y = (12/0.03) = 400$. The line itself is illustrated in Figure 2.10. This line also does not intersect the feasible region. The implementation of the isocost line process must be repeated so that after each repetition a new isocost line appears parallel to the previous ones. The slope of all the lines is determined by the coefficients of x and y, which are $(0.02, 0.03)$, and does not change from one isocost line to another. An interesting fact in this case is that the farther the isocost line is from the origin, the higher the cost associated with the isocost line is. (This is because the cost is proportional to the positive values, namely, 0.02 and 0.03.) Therefore, the goal of this example is to discontinue the process at the lowest possible cost or at the line that is located as close as possible to the origin. Practically speaking, the next step is to take a line and move it parallel to one of the isocost lines until it touches the feasible region for the first time. The first contact point is the optimal one; this point is clearly indicated in Figure 2.10. It is obvious that moving farther inside the feasible region does, in fact, produce other feasible points, but these are associated with higher cost, since the points are situated on isocost lines that are located farther from the origin. Typically, the coordinates of the optimal point can be easily read from the graph or in other instances can be precisely calculated from the equations. In this instance, the optimal point lies on the intersection of the lines that indicate vitamin A and vitamin E as the boundaries. Therefore, it is apparent that the optimal point is the solution of the following equations:

$$\text{vitamin A} \quad 0.1x + 0.05y = 40$$
$$\text{vitamin E} \quad 0.06x + 0.1y = 60$$

or

$$0.2x + 0.1y = 80$$
$$0.06x + 0.1y = 60$$

Subtraction of the second equation from the first one yields

$$0.14x = 20$$

$$x = \frac{20}{0.14} = \frac{1000}{7} \qquad y = \frac{60 - 0.06x}{0.1} = \frac{3600}{7}$$

From the above discussion, it should be inferred that the optimal solution is

$$x = \frac{1000}{7} \quad \text{and} \quad y = \frac{3600}{7} \text{ units}$$

and by substituting the x and y values, respectively, in the objective function, we obtain the total cost:

$$c = 0.02x + 0.03y$$
$$= 0.02 \cdot \frac{1000}{7} + 0.03 \cdot \frac{3600}{7}$$
$$= \frac{20}{7} + \frac{108}{7} = \frac{128}{7} \quad \text{or} \quad \$18.29$$

which is the cost associated with the problem.

It is interesting to observe the way in which this solution satisfies all five constraints. The amount of vitamin A purchased by the vegetarian is

$$0.10x + 0.05y = 0.10 \cdot \frac{1000}{7} + 0.05 \cdot \frac{3600}{7}$$
$$= \frac{100}{7} + \frac{180}{7} = \frac{280}{7} \quad \text{or} \quad 40 \text{ grams}$$

which is the minimal requirement for the diet.

For vitamin B, the solution is

$$0.20x + 0.15y = 0.20 \cdot \frac{1000}{7} + 0.15 \cdot \frac{3600}{7}$$
$$= \frac{200}{7} + \frac{540}{7} = \frac{740}{7} \quad \text{or} \quad 105.71 \text{ grams}$$

which is more than the required 50 grams. The difference,

$$\frac{740}{7}-50=\frac{390}{7} \quad \text{or} \quad 55.7 \text{ grams}$$

is the distance OB marked in Figure 2.11, between the optimal solution and the constraint B.

The amount of vitamin C purchased is

$$0.04x+0.20y=0.04\cdot\frac{1000}{7}+0.20\cdot\frac{3600}{7}$$

$$=\frac{40}{7}+\frac{720}{7}=\frac{760}{7} \quad \text{or} \quad 108.57 \text{ grams}$$

which is more than the required 70 grams.

The difference,

$$\frac{760}{7}-70=\frac{270}{7} \quad \text{or} \quad 38.57 \text{ grams}$$

is the distance OC marked in Figure 2.11 between the optimal solution and constraint C.

For vitamin D, the optimal solution yields

$$0.10x+0.10y=0.10\cdot\frac{1000}{7}+0.10\cdot\frac{3600}{7}$$

$$=\frac{100}{7}+\frac{360}{7}=\frac{460}{7} \quad \text{or} \quad 65.71 \text{ grams}$$

which is more than the 10 grams required for the diet.

Figure 2.11. The Diet Problem—Distances from Optimal to Required

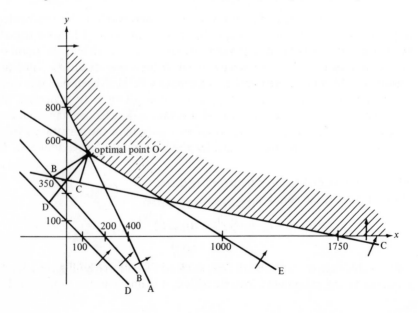

The difference,

$$\frac{460}{7} - 10 = \frac{390}{7} \quad \text{or} \quad 55.71 \text{ grams}$$

is the distance OD between the optimal solution and the constraint D, and is also indicated in Figure 2.11.

The quantity of vitamin E that the shopper selects is

$$0.06x + 0.10y = 0.06 \cdot \frac{1000}{7} + 0.10 \cdot \frac{3600}{7}$$

$$= \frac{60}{7} + \frac{360}{7} = \frac{420}{7} \quad \text{or} \quad 60.00 \text{ grams}$$

which is the precise amount specified by the diet. At the optimal solution, the restrictions on vitamins A and E are active whereas those on B, C, and D are inactive.

2.5. Special Situations of Graphical Solutions

The examples of the Exclusive Furniture Company and the vegetarian problem in this chapter have provided an opportunity to examine typical linear programming problems. This section contains examples of special cases of linear programming problems. The special cases are neither irregular nor unusual from a mathematical point of view; however, poor model building, or in other instances mistakes in the input data, could lead to these special cases.

2.5.1. Unbounded Feasible Region

As indicated in the vegetarian problem, it is conceivable to be confronted with a linear programming problem that has an unbounded feasible region. In this case, it is possible that the optimal solution is on an extreme point of a feasible region that is unbounded from at least one direction. That the feasible region is not closed and is not properly defined from one side does not prevent us from searching for the optimal solution. In reviewing the facts, bear in mind that the optimal solution is located on the bounded feasible region. It is important to focus on a portion of the constraints from Section 2.4. In this problem, consider the constraints for vitamins A, C, and E:

$$0.10x + 0.05y \geqslant 40$$
$$0.04x + 0.20y \geqslant 70 \qquad (2.5)$$
$$0.06x + 0.10y \geqslant 60$$
$$x \geqslant 0 \quad y \geqslant 0$$

The feasible region for Problem (2.5) is clearly illustrated in Figure 2.12. In reference to the unbounded feasible region, it is important to point out that

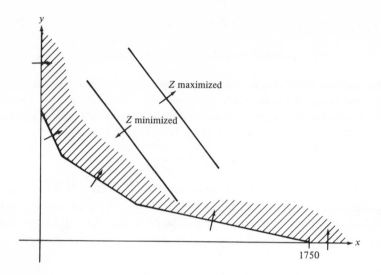

Figure 2.12. The Diet Problem—Feasible Region for Vitamins A, C, and E

two distinct cases are associated with this example: The first case occurs when the objective function generates a finite value, and therefore the optimal solution is finite. A good illustration of this case is given in the vegetarian problem, where the objective function is stated as

$$\text{minimize } 0.02x + 0.03y$$

and the solution is

$$x = \frac{1000}{7} \qquad y = \frac{3600}{7}$$

and the cost has been determined to be $18.29.

On the other hand, in the second (the unusual) case, the objective function is

$$\text{maximize } 5x + 4y$$

In relation to the objective function, there is no upper limit for the value of the optimal solution. In fact, $x = \infty$, $y = \infty$ is a feasible solution and the value of the objective function is infinitely high. As indicated, this case is unusual because in real-life situations it is impossible for the decision variables to become infinite and the objective function to reach infinity. Thus, in most cases, given the previous facts, it is safe to assume that a mistake was made either in the building of the model or in the input data.

There is yet another case, often regarded as being "in between" the two previous cases. For the same feasible region, let us assume that the objective function is

$$\text{minimize } c_1 x + c_2 y.$$

For $c_1 = 0$, $c_2 = 1$, the objective function is

$$\text{minimize } (0x + 1y) = \text{minimize } y.$$

Apparently, an optimal solution for this problem is $y = 0$, $x = 1750$ (since the first point in the feasible region to touch the x axis is $(1750, 0)$) and the value of the objective function is 0. In many ways, this example is similar to the first case mentioned. There is, however, another solution,

$$y = 0, \qquad x = \infty$$

in which case the value of the objective function is equal to 0. Since x has an infinite value, it is unlikely that a real-life situation would call for such a solution. Therefore, a review is recommended in order to trace possible logical mistakes in the model.

2.5.2. Infeasible Region

In all instances, the optimal solution must be a member of the feasible region. As pointed out before, the feasible region is the intersection of all constraints. It is possible, however, that a few of the constraints may contradict one another. One implication is then clear: There is no solution that will simultaneously satisfy all the constraints.

Let us consider the following problem as a means of understanding this situation. If one vice-president imposes a restriction that no more than 25 man-hours can be spent on a project, and another vice-president demands that at least 35 man-hours be allocated to the same task, it is easy to conclude that it is impossible to satisfy both executives.

The geometric meaning of an infeasible solution can be appreciated by observing the intersection of all the constraints. If this area is the empty set (the null set), that is, if there is no region that is common to all constraints, then no feasible solution exists for the model. The following example simplifies the discussion. The two constraints are

$$2x + 3y \leqslant 25$$
$$4x + 6y \geqslant 70$$

By looking at Figure 2.13, we can see that only region I satisfies the first constraint, whereas only region II satisfies the second constraint. It is apparent that no point exists that satisfies both constraints at the same time. An important circumstance that may lead to an infeasible solution is the reversal of an inequality sign in the process of finding a solution. For example, the sign greater than or equal to ($\geqslant$) may have been written instead of less than or equal to ($\leqslant$), and vice versa.

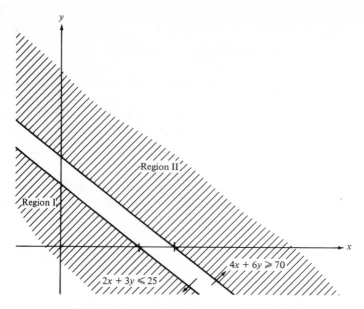

Figure 2.13. Example of No Feasible Region

2.5.3. One-Dimensional Linear Programming

A degenerate case of a linear programming problem occurs when there is only one decision variable. Such problems are referred to as one-dimensional linear programming. In order to illustrate this case, we present the following example.

The BL toy company, located in Penn Valley, has the production capacity to manufacture only one toy. This item requires certain amounts of plastic and metal as basic raw materials necessary for production. In each toy that is manufactured there are 20 grams of plastic and 40 grams of metal. The daily allowances for plastic and metal are 5 and 8 kilograms, respectively. The assembly process is completed on a machine that is available 10 hours per day, and time and motion studies indicate that it takes 10 minutes to assemble one toy. There is another important fact to be considered: To ensure maximum operational performance it is necessary to operate the machine for at least 5 hours a day. In addition, the marketing reports indicate that the company must produce at least 20 toys a day to satisfy the retailers' demands. If the company has a policy of maximizing its profit at a price of $7 a toy, determine the optimal number of toys that the manufacturer should produce each day in order to maximize profits.

This case represents a one-dimensional linear programming problem in the sense that the only decision variable is the number of toys to be produced, which is represented by x.

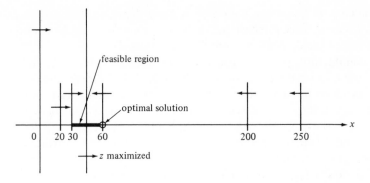

Figure 2.14. BL Toy Company—One-Dimensional Linear Programming

Constraints	Restrictions pertaining to
$20x \leqslant 5000$	plastic
$40x \leqslant 8000$	metal
$10x \leqslant 10 \cdot 60$	machine— 10 hours
$10x \geqslant 5 \cdot 60$	machine— 5 hours
$x \geqslant 20$	previous contract
$x \geqslant 0$	nonnegative restriction
maximize $7x$	objective function

From Figure 2.14 it is evident that this illustration contains only one dimension, the x axis. The first constraint to consider is the plastic material usage. The restriction on x is listed as less than or equal to 250. The second constraint, on the metal resource, restricts x to be no more than 200. In addition, the machine time usage constraints limit the number of toys to be no more than 60, but no less than 30. Finally, a previous contract restricts x to be at least 20. Thus, we may generalize that the intersection of all six constraints, including $x \geqslant 0$, is $30 \leqslant x \leqslant 60$. It is evident that this area is also the feasible region for the problem. It must be noted, however, that there are two extreme points for the feasible region, $x = 30$ and $x = 60$, and the objective function $7x$ increases while moving from left to right. The facts indicate that the optimal solution is $x = 60$, and the profit derived from the manufacturing process is $\$7 \cdot 60 = \$420$.

2.5.4. Three-Dimensional Linear Programming

By now you are aware of the advantages and disadvantages of the graphical method of solving linear programming problems. The graphical method can be easily applied as long as the number of decision variables is either one or two. The purpose of this section is to introduce some of the difficulties associated with linear programming in more than two dimensions. Specifically, a typical three-dimensional linear programming problem is presented and some of the difficulties it entails are emphasized.

A manufacturer produces calculators in three models, A, B, and C. The profit per unit on the calculators is \$10, \$15, and \$30, respectively. Each of these products is scheduled to be processed in three departments—the printing department, the assembly line, and the quality control department. The time required for the calculators A, B, and C at the printing department is 2, 3, and 3 hours, respectively. The production schedule further indicates that the products must be then processed on the assembly line for 0.5, 0.8, and 1.2 hours, respectively. Finally, each different model will require 0.5 hour in the quality control department. However, there are restrictions on the available hours in the three departments. There are only 15 available hours per day in the printing department, 10 available hours per day on the assembly line, and 5 hours per day available in the quality control department. The decision to be made is how many of each model to produce each day in order to maximize profit.

Let x, y, and z represent the number of calculators for types A, B, and C, respectively. The objective function is represented as

$$\text{maximize } 10x + 15y + 30z$$

Since there are only 15 hours available in the printing department, the first constraint is

$$2x + 3y + 3z \leqslant 15$$

There is also a time restriction for the assembly line, which leads to the next constraint:

$$0.5x + 0.8y + 1.2z \leqslant 10$$

Finally, the restriction for the quality control department is

$$0.5x + 0.5y + 0.5z \leqslant 5$$

with $x \geqslant 0$, $y \geqslant 0$, $z \geqslant 0$; thus the problem is

$$
\begin{aligned}
\text{Maximize} \quad & 10x + 15y + 30z \\
\text{subject to} \quad & 2x + 3y + 3z \leqslant 15 \\
& 0.5x + 0.8y + 1.2z \leqslant 10 \qquad (2.6) \\
& 0.5x + 0.5y + 0.5z \leqslant 5 \\
\text{and} \quad & \\
& x \geqslant 0 \qquad y \geqslant 0 \qquad z \geqslant 0
\end{aligned}
$$

Let us now use the graphical method to solve Problem (2.6). The decision variables associated with the problem are x, y, and z. Therefore, the solution is a point in the three-dimensional space. The three constraints are plotted in the x, y, z space and the feasible region is obtained as indicated in Figure 2.15.

Convert the first constraint of Problem (2.6)

$$2x+3y+3z \leqslant 15$$

to an equality and begin the search for a solution with the equation $2x+3y+3z=15$. The equality intersects the three axes at these points:

$$\text{when } y=0 \text{ and } z=0 \text{ then } x=\frac{15}{2}=7.5$$

$$\text{when } x=0 \text{ and } z=0 \text{ then } y=\frac{15}{3}=5$$

$$\text{when } x=0 \text{ and } y=0 \text{ then } z=\frac{15}{3}=5$$

and the plane passes through the three points $(7.5, 0, 0)$, $(0, 5, 0)$, and $(0, 0, 5)$.

Since the first constraint is an inequality, one of the half spaces is feasible and the other half is infeasible. Bear in mind that it is necessary to verify only one of the points.

Figure 2.15. Three-Dimensional Linear Programming

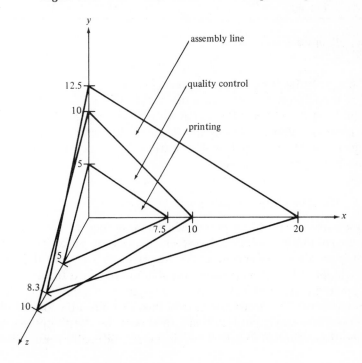

Let us proceed to the next step and consider the origin $(0,0,0)$. It is evident that this point satisfies the constraint $2 \cdot 0 + 3 \cdot 0 + 3 \cdot 0 \leqslant 15$ and therefore the half space containing the origin is the feasible region.

Since the variables must be nonnegative, the feasible region is located inside the pyramid formed by the four coordinates $(7.5,0,0)$, $(0,5,0)$, $(0,0,5)$, and $(0,0,0)$.

The two restrictions concerning the assembly line and the quality control departments form two different pyramids with the coordinates $(20,0,0)$, $(0,12.5,0)$, $(0,0,8.33)$, and $(0,0,0)$ for the assembly line and $(10,0,0)$, $(0,10,0)$, $(0,0,10)$, and $(0,0,0)$ for the quality control department. The feasible region is located at the intersection of the three pyramids. Fortunately, there is an easy solution to this problem because the intersection is the pyramid associated with the printing department. The objective function is not plotted deliberately, instead, the extreme points of the feasible region will be evaluated.

The extreme points of the feasible region are the four corners of the pyramids.

The value of the objective function at the various points are

At point	the value of the objective function is
$(7.5,0,0)$	$10 \cdot 7.5 + 15 \cdot 0 + 30 \cdot 0 = \$75$
$(0,5,0)$	$10 \cdot 0 + 15 \cdot 5 + 30 \cdot 0 = \$75$
$(0,0,5)$	$10 \cdot 0 + 15 \cdot 0 + 30 \cdot 5 = \$150$
$(0,0,0)$	$10 \cdot 0 + 15 \cdot 0 + 30 \cdot 0 = \$0$

Thus, the optimal solution is

$$x=0 \qquad y=0 \qquad z=5$$

and the profit for the production process is \$150.

You should now be aware of some of the difficulties involved in solving this problem if you begin the search with the extreme points formed by the two pyramids associated with the assembly line and the quality control departments. Most of the trouble is attributable to plotting the intersection of planes in three dimensions.

2.5.5. Multiple Optimal Solutions

All the numerical problems presented thus far have had a common property —a unique optimal solution. However, the decision maker working with a real-life problem may encounter conditions that have more than one optimal solution. While implementing the solution obtained from a mathematical programming problem, it is advantageous to know that alternative solutions exist. The decision maker is then in a position to use subjective factors—perspective and intuition—to choose the "best" optimal solution among these alternatives. In other instances, difficulties may arise when some of the constraints are not mathematically formulated, causing these constraints not to appear in the model. For example, a particularly difficult situation can occur if the solution is sensitive to changes in one resource; in this case the decision maker may prefer to use a more stable and comfortable alternative.

The following is a typical case in which more than one optimal solution may be available to the decision maker. Farmer Izab has 200 acres and grows two vegetables— tomatoes and potatoes. An acre of potatoes requires 2 gallons of water a day, and an acre of tomatoes requires 3 gallons of water a day. The total amount of water available is restricted to 500 gallons per day. In addition, each acre of potatoes requires 3 man-hours per day, and each acre of tomatoes requires 1 man-hour per day. The farmer is further restricted to 250 man-hours every day. Each acre of potatoes contributes $400 to profit and each acre of tomatoes contributes $600 to profit. The decision to be made is how many acres of tomatoes and potatoes to plant in order to maximize profits.

Let x and y represent the number of acres for potatoes and tomatoes, respectively. Since only 200 acres are available for the two vegetables, the first constraint is

$$x+y \leqslant 200$$

The water supply is restricted to 500 gallons a day, which is indicated in the constraint

$$2x+3y \leqslant 500$$

There is also a time limitation; thus the man-hour restriction is

$$3x+y \leqslant 250$$

The objective function is to maximize $400x+600y$ and the variables must be nonnegative $x \geqslant 0, y \geqslant 0$. The mathematical programming problem can be stated as follows:

$$
\begin{array}{lll}
\text{Maximize} & 400x+600y & \\
\text{subject to} & x+y \leqslant 200 & \\
& 2x+3y \leqslant 500 & \quad (2.7) \\
& 3x+y \leqslant 250 & \\
\text{and} & & \\
& x \geqslant 0 \quad y \geqslant 0 &
\end{array}
$$

The graphical solution appears in Figure 2.16.

The objective function is maximized in the direction as indicated in Figure 2.16. It is clear that the objective function reaches the feasible region at a line and not a point. The segment AB is the optimal solution, and in fact any point on the segment AB should be considered an optimal solution.

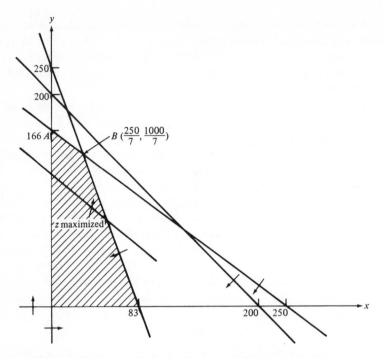

Figure 2.16. Farmer Izab Problem—Multiple Optimal Solutions

The coordinate of A is $(0, 166.66)$ and the coordinate of B is $(250/7, 1000/7)$. Let us extend this problem a little further and choose an arbitrary point C $(25, 150)$ that is located between A and B. The value of the objective function at the three points may be listed as

point A	$\$400 \cdot 0 + \$600 \cdot 166.66 = \$100,000$
point B	$\$400 \cdot 250/7 + \$600 \cdot 1000/7 = \$100,000$
point C	$\$400 \cdot 25 + 600 \cdot 150 = \$100,000$

The solutions clearly indicate that there is a profit of $100,000 at the three points A, B, and C. You can further verify that any point between A and B also has to have the property of earning $100,000.

Farmer Izab has to choose the optimal solution, since he has many alternatives. The next step in this process is to consider other facts that may influence the decision. For example, if he does not want to lose his reputation in the market as a potato farmer, he should choose point B with $x = 250/7$ for potatoes and $y = 1000/7$ for tomatoes. However, if he decides to concentrate his efforts on growing tomatoes, then it would be wise for him to choose point A, which indicates no potatoes and $y = 166.66$ acres for tomatoes. Thus the farmer must make a subjective decision, since both solutions contribute a $100,000 profit.

2.6. Summary

This chapter presents the general linear programming problem. It shows that many different types of problems can be *mathematically modeled* into a form appropriate for a linear program. The specific requirements for a linear program are that there exist *constraints* and an *objective function* that are linear, and that the *variables* satisfy the *nonnegativity restriction*.

A first step toward solving linear programming problems, a *graphical solution* method, is demonstrated. Though restricted in applicability by the difficulties associated with graphing in three or more dimensions, this method clarifies certain basic properties of linear programs. On a graph it is easy to distinguish points that satisfy all of the constraints, including the nonnegativity restrictions, and to group these points into a single *feasible region*. The feasible region is bounded by some or all of the constraint lines, and the *intersection points* that are feasible form the set of *extreme points*. If a solution exists then there is a solution at an extreme point, since this is where the *isoprofit line* or isocost line last touches the feasible region.

Special cases are considered. One is when the isoprofit line is parallel to the constraint line farthest from the origin and yields *multiple optimal solutions*. A second special case obtains when there is no limit to the feasible region in one direction and this direction is the one in which the isoprofit line should be moved, so that the solution is *unbounded*. The last special case is one in which the constraints are inconsistent and thus *no feasible solution* exists. It is also observed that some constraints may not be considered in the problem because one or more of the other constraints make them *redundant*. On the other hand, those constraints that intersect at an extreme point limit the direction in which movement can be made and thus are said to be *active* at that point.

Since the graphical solution method is basically useful only for two-dimensional problems, it is necessary to develop a different method for solving problems with three or more variables. The one described in the next chapter serves this purpose. However, it should be borne in mind that the solution method presented in Chapter 3 is simply the algebraic representation of the graphical method of this chapter.

References and Selected Readings

Beale, E. M. L. 1968. *Mathematical Programming in Practice*. Belmont, California: Fearon-Pitman.

Dallenbach, H. G., and E. J. Bell 1970. *User's Guide to Linear Programming*. Englewood Cliffs, New Jersey: Prentice-Hall.

Driebeck, N. J. 1969. *Applied Linear Programming*. Reading, Massachusetts: Addison-Wesley.

Kreko, B. 1968. *Linear Programming*. New York: American Elsevier.

Martin, E. W., Jr., and R. H. Hermanson 1969. *Linear Programming*. Homewood, Illinois: Irwin.

Strum, Jay E. 1972. *Introduction to Linear Programming*. San Francisco, California: Holden-Day, Inc.

Stockton, R. S. 1971. *Introduction to Linear Programming*. Homewood, Illinois: Irwin.

Throsby, C. D. 1970. *Elementary Linear Programming*. New York: Random House.

Vajda, S. 1961. *Mathematical Programming*. Reading, Massachusetts: Addison-Wesley.

Wolfe, C. S. 1970. *Linear Programming with Fortran*. Glenview, Illinois: Scott, Foresman & Co.

Problems

Formulate Problems 1 Through 8 as Linear Programming Problems

1. The Battel toy manufacturing company produces two types of puppets, the "Teddy" model and the "Jimmy" model. The assembly department requires 6 minutes of machine time for each Teddy and 10 minutes for each Jimmy model. In the finishing department 2 minutes of labor are needed for each Teddy and 4 minutes for each Jimmy. (Note that Jimmy needs labor more than Teddy does). Finally, the in-house quality control department takes 4 minutes of labor to inspect a puppet of either kind.

   There are four machines available in the assembly department, each of which operates 45 minutes per hour. Four and two men work in the finishing and quality control sections, respectively. Time studies indicate that workers rest approximately 10 minutes per each hour. If each Teddy doll contributes $50.00 to profit and each Jimmy doll contributes $60.00, find the optimal hourly production of each type of puppet.

2. The Temple Company must meet demand requirements of 81, 48, and 45 units for products A, B, and C, respectively. These units can be manufactured by using either of two general-purpose machines. However, the machines' production capabilities as well as their operation costs differ, as shown in the table.

Machine	Product			Cost, cents per hour
	A	B	C	
M-1	9	8	15	12
M-2	27	12	9	28

The entries are expressed in number of units per hour that can be made on each machine. Find the production schedule that minimizes cost.

3. Certain vegetarians eat only salad and bread provided that they consume 70, 86, and 62 units, respectively, of vitamins A, B, and C per day. A pound of salad contains 7 units of vitamin A, 4 units of vitamin B, and 3 units of vitamin C. A loaf of bread contains 2 units of A, 4 units of B, and 2 units of C.

   The cost of bread is 50 cents per loaf and the cost of salad is 10 cents a pound. Find the combination of bread and salad that these vegetarians should eat daily to minimize cost.

4. Agatha Christin is the library manager of Presley University. With a budget of $1400 she is considering the purchase of book stacks to cover 720 square feet of floor space. There are basically two models in the market; Model A has a storage capacity of 80 cubic feet, requires 60 square feet of floor space, and costs $100 per book stack; Model B stores 120 cubic feet of books, requires 80 square feet, and costs $200 per book stack. How many book stacks of each model should she buy in order to maximize storage capacity?

5. North Amsterdam publishing company is releasing three new books next month. Four raw materials are used in the production process, but their availability for the next month is restricted as follows: 1000 pounds of paper, 80 pounds of ink, 2000 machine-hours, and 4000 man-hours.

   Contractual commitments to the authors require that at least 50 copies of the first book and 40 copies each of the second and third books be printed. The table shows the use of raw materials, measured in pounds per copy and hours per copy for each book. Book 1 will sell for $15 a copy, book 2 for $20, and book 3 for $16.

	Book		
	1	2	3
Paper	10	15	18
Ink	0.50	0.80	1.00
Machine–hours	30	24	20
Man–hours	50	45	46

   Assume that all copies that are printed will eventually be sold. Determine the number of copies of each book to be printed to maximize revenues.

6. David Lass is the production manager at Startek Enterprises. He would like to determine which production mix of three products, A, B, and C, should be scheduled for the next month. There are only 6000 units of materials and 3680 man-hours available for the month. Product A uses 4 units of materials and 4 man-hours; Product B uses 48 units of materials and 32 man-hours; and Product C needs 60 units of materials and 24 man-hours. The contribution to profit is $7.00, $8.50, and $9.00 for Product A, B, and C, respectively. How many units of each product should David manufacture?

7. Vel Company manufactures and distributes one product in five regions of the United States. Recently, a new plant was opened and will begin to operate this

month. The company owns a warehouse in each of the five regions, which are supplied by both the old and the new plant.

The owner, Sam Vel, wants to know how to allocate this month's production between the plants. He is concerned with two types of costs: first, the distribution costs incurred when shipping the finished product to the warehouses; second, the production costs at the plants, which are expected to be lower at the new plant due to a more advanced technology. He also faces two restrictions, the resource availability and the orders placed by the warehouses.

The tables show the required data.

Table I

	Old plant	New Plant	Limit on Resources
Production cost per unit	$3.00	$2.50	
Man-hours/unit	10	8	12,000
Units of materials/ unit	18	9	24,000

Table II Shipping Costs (in dollars per unit)

	Regions				
	A	B	C	D	E
Old plant	1.30	2.40	4.40	1.50	5.30
New plant	4.10	0.50	2.30	1.00	2.80
Demand requirements in units	80	100	200	300	150

8. Arnie Arco has been offered the chance to participate in two business ventures. One venture calls for a $3000 payment and 10 hours of work a week for each unit (share) of participation. The second one requires $5000 a unit of participation and 5 hours of work a week for each unit. Arnie has only $15,000 available and 25 hours a week of free time. If the first venture returns a net profit of $8000 per year and the second returns $6000 per year, find the optimal number of units of participation for each venture (assuming fractional units are permitted).

9. Solve Problem 1 using the graphical method.

10. Solve Problem 3 using the graphical method.

11. Solve Problem 4 using the graphical method.

12. Solve Problem 8 using the graphical method.

Solve Problems 13 *Through* 21 *Using the Graphical Method*

13. Maximize $z = 4x + 6y$

    subject to $x + y \leqslant 8$

    $\qquad 3.5x + 2y \geqslant 14$

    $\qquad\qquad x \leqslant 4$

    $\qquad\qquad y \leqslant 7$

    $\qquad x \geqslant 0 \qquad y \geqslant 0$

14. Minimize $z = 24x + 15y$

    subject to $y - x \geqslant 0$

    $\qquad 7x + 11y \geqslant 77$

    $\qquad 16x + 8y \geqslant 80$

    $\qquad x \geqslant 0 \qquad y \geqslant 0$

15. Maximize $z = x + y$

    subject to $6x + 10y \leqslant 120$

    $\qquad\qquad 5 \leqslant x \leqslant 10$

    $\qquad\qquad 3 \leqslant y \leqslant 8$

    $\qquad x \geqslant 0 \qquad y \geqslant 0$

16. Minimize $z = 10x + 14y$

    subject to $20x + 5y \geqslant 60$

    $\qquad\qquad 8x + 6y \geqslant 48$

    $\qquad\qquad 4x + 8y \geqslant 40$

    $\qquad x \geqslant 0 \qquad y \geqslant 0$

17. Maximize $z = 6x + 9y$

    subject to $6x + 3y \leqslant 36$

    $\qquad\qquad 2x + 3y \leqslant 24$

    $\qquad x \geqslant 0 \qquad y \geqslant 0$

18. Minimize $z = 3x + 2y$

    subject to $15y - 9x \geqslant 60$

    $\qquad\qquad 6y + 18x \leqslant 54$

    $\qquad x \geqslant 0 \qquad y \geqslant 0$

19. Minimize $z = 7x - 5y$

    subject to $11x - 10y \geqslant -20$

    $\qquad\qquad 0 \leqslant y \leqslant 6$

    $\qquad\qquad x \geqslant 0$

20. Maximize $z = x + 2y$

subject to $2y - x \geqslant 12$

$$y - 2x \geqslant 3$$

$$x \geqslant 0 \quad y \geqslant 0$$

21. Minimize $z = 10x + 2y$

subject to $x + y \geqslant 1$

$$y - 5x \leqslant 0$$

$$x - y \geqslant -1$$

$$x, y \geqslant 0$$

22. For the following constraints:

$$2x + 2y \leqslant 6$$

$$x - 2y \leqslant 4$$

$$0 \leqslant x \leqslant 4$$

$$0 \leqslant y \leqslant 1$$

(a) List all intersection points.

(b) List all extreme points.

(c) Calculate the value of the function $15x + 10y$ at all extreme points.

23. **(a)** What is the maximum number of intersection points if there are n variables and m constraints?

(b) Can the actual number of intersection points be less than in part (a)? Why?

24. Solve each of the following linear programming problems by the graphical method.

(a) $2x + y \leqslant 15$

$$x + y \geqslant 3$$

$$x - 2y \leqslant 8$$

$$x \geqslant 0 \quad y \geqslant 0$$

Objective functions: Maximize $x + 2y$; maximize x; minimize y;
minimize $-x - 2y$.

(b) $3x + 4y \leqslant 20$

$$x - y \leqslant -3$$

$$4 \leqslant x + 2y \leqslant 8$$

$$x, y \geqslant 0$$

Objective functions: Maximize $3x + y$; maximize $3x + 4y$;
minimize $3x + 4y$.

(c) $x + y + z \leqslant 5$

$$x + y \geqslant 0$$

$$x, y, z \geqslant 0$$

Objective functions: Maximize $x + y$; minimize $y + z$; maximize $x + y + z$.

Three

Linear Programming— The Simplex Method

3.1. Introduction

The second chapter, presented a graphical method that can be used to solve a linear programming efficiently. However, the limitations of that method can be serious, since with it we can solve only for two or three variables. Needless to say, few real-life problems contain only two decision variables; in fact, some situations involve thousands of decision variables and hundreds of constraints. Therefore, in this chapter we present an algorithm that can be used to solve linear programming problems of any size. This algorithm is an iterative computational technique called the *simplex method*.

Recall that an algorithm is a set of rules for solving a problem that frequently involves the repetition of a procedure. An iterative algorithm is one that includes many such cycles. Each cycle is similar to the preceding one in terms of the rules employed, and at the end of each cycle there is an improved solution. Thus, after each iteration the solution is closer to optimality. (An example was given in Section 1.4.)

The first algorithm for linear programming was developed in 1947 by Dantzig (1963). Since then, many modifications have been suggested, basically in order to reduce the computational effort involved. The improvements are also intended to decrease the storage requirements for computerized versions of the algorithm. The most advantageous characteristic of the simplex method is the simplicity of its calculation. In this respect, the process can be easily programmed for any computer. Specifically, the algorithm requires only addition, subtraction, multiplication, and division.

3.2. The Simplex Computational Procedure

A typical linear programming problem is a maximization problem that can be formulated as

$$\text{Maximize} \qquad z = c_1 x_1 + c_2 x_2 + \cdots + c_n x_n$$

$$\text{subject to} \qquad a_{11} x_1 + a_{12} x_2 + \cdots + a_{1n} x_n \leqslant b_1$$

$$a_{21} x_1 + a_{22} x_2 + \cdots + a_{2n} x_n \leqslant b_2 \qquad (3.1)$$

$$\vdots$$

$$a_{m1} x_1 + a_{m2} x_2 + \cdots + a_{mn} x_n \leqslant b_m$$

$$x_j \geqslant 0 \qquad j = 1, 2, \ldots, n$$

There are n variables and m constraints.

A *basic solution* to Problem (3.1) is one that contains exactly m positive variables. (The other $n - m$ variables are set equal to zero.)

Two basic solutions are *adjacent* to one another if exactly $m - 1$ of their m positive variables are the same variables and only one of their positive variables differs. Geometrically this means that the two basic solutions are two extreme points connected by a straight line, and that it is possible to "slide" from one basic solution to its adjacent extreme point along that straight line.

The procedure is based on the main theorem of linear programming, which associates an extreme point to a basic solution (see Gass, 1975, page 55). It was concluded in Chapter 2 that if there is an optimal solution to a linear programming problem, then it must be at one of the extreme points. This conclusion can now be extended: *If there is a solution to a linear programming problem, it must be one of the basic solutions.* This enlightening statement means that it is sufficient to search for the optimal solution among the basic solutions. In effect, this statement summarizes the simplex procedure, which provides for a systematic search among the basic solutions. If the current basic solution is not an optimal one, the procedure generates an adjacent basic solution that is better than the preceding one. Finally, after a finite number of iterations, an optimal solution is reached if it exists. The four steps of the simplex algorithm follow.

1. Preparation of the problem. The preparation phase is the conversion of the linear programming problem into a *canonical*, or standard, form. A canonical form is one in which the right-hand side of the constraints (sometimes known as RHS or the b_i) is nonnegative. Thus, if one of the b_i is negative, it is necessary to multiply that equation by -1. In addition, it is important that all the constraints be in an equality form. In order to achieve this form, it is necessary to add nonnegative values, called *slack variables s_i,* to the smaller sides of the inequalities. These slack variables balance the actual use of the left-hand sides of the inequalities with the availability of the right-hand sides. Thus, Problem (3.1) can now be written as follows:

Maximize

$$z = c_1 x_1 + c_2 x_2 + \cdots + c_n x_n$$

subject to

$$a_{11} x_1 + a_{12} x_2 + \cdots + a_{1n} x_n + s_1 = b_1$$

$$a_{21} x_1 + a_{22} x_2 + \cdots + a_{2n} x_n + s_2 = b_2 \qquad (3.2)$$

$$\vdots \qquad \vdots \quad \vdots \qquad \vdots \qquad \vdots$$

$$a_{m1} x_1 + a_{m2} x_2 + \cdots + a_{mn} x_n + s_m = b_m$$

$$x_j \geqslant 0 \quad j = 1, 2, \ldots, n \qquad s_i \geqslant 0 \quad i = 1, 2, \ldots, m \qquad b_i \geqslant 0 \quad i = 1, 2, \ldots, m$$

One implication is now clear: The slack variables do not appear in the objective function. That there is some slack s_i does not affect the value of z of the objective function. In this respect, the coefficients of s_i are zero in the objective function.

2. *Starting solution.* An easy starting solution is

$$x_j = 0 \quad j = 1, 2, \ldots, n \qquad s_i = b_i \qquad i = 1, 2, \ldots, m$$

It should be recognized, however, that it is not always easy to find a starting solution in linear programming problems. In cases where an easy solution does not exist, an artificial one is generated (see Section 3.5). The positive variables are called basic variables and in this case are also the slack variables. The nonbasic variables assume the value of 0 and are the original decision variables x_j. The value of the objective function is $z = 0$.

To facilitate discussion, we now present the linear programming problem in tableau form. Several formats are used but the basic premise is the same. Table 3.1 contains the starting tableau.

Each row in Table 3.1 is an equation of Problem (3.2). The basic variables are the first starting solution and are listed in the basic variables column. Above the list of variables are the coefficients of the variables in the objective function; these coefficients are denoted by c_j. Adjoining the basic variables column is another column, headed c_b, of the coefficients of the basic variables in the objective function. In order to generate the last row it is necessary to multiply the c_b column by each column and to subtract the coefficient of that column. Consider the following example:

The first column is

$$0 \cdot a_{11} + 0 \cdot a_{21} + \cdots + 0 \cdot a_{m1} - c_1 = -c_1$$

and for the second column

$$0 \cdot a_{12} + 0 \cdot a_{22} + \cdots + 0 \cdot a_{m2} - c_2 = -c_2$$

For the value column, multiply the coefficient c_b by the values to obtain

$$0 \cdot b_1 + 0 \cdot b_2 + \cdots + 0 \cdot b_m = 0$$

Table 3.1 The Simplex Tableau

c_b	Basic variables	c_j Value	c_1 x_1	c_2 x_2	$\cdots$	c_n x_n	0 s_1	0 s_2	$\cdots$	0 s_m
0	s_1	b_1	a_{11}	a_{12}	$\cdots$	a_{1n}	1	0	$\cdots$	0
0	s_2	b_2	a_{21}	a_{22}	$\cdots$	a_{2n}	0	1	$\cdots$	0
$\vdots$	$\vdots$	$\vdots$	$\vdots$							
0	s_m	b_m	a_{m1}	a_{m2}	$\cdots$	a_{mn}	0	0	$\cdots$	1
		0	$-c_1$	$-c_2$	$\cdots$	$-c_n$	0	0	$\cdots$	0

It should be emphasized that the value 0 is the current solution of the objective function. Generally speaking, it is quite simple to read the solution from Table 3.1. Furthermore, the information which is necessary to solve the problem is stored in a table that is $(m+3)\times(n+m+3)$. The current solution is given by the variables listed in the basic variables column and their values appearing in the value column. Moreover, all other variables in the problem that are not in the basis, that is, variables which do not appear in the column headed "Basic variables", have the value 0. Therefore, a solution to the problem, though not necessarily the optimal one, is

$$s_i = b_i \qquad i = 1, 2, \ldots, m$$

$$x_j = 0 \qquad j = 1, 2, \ldots, n$$

$$z = 0 \qquad \text{the value of the objective function}$$

The coefficients of the last row with reversed signs, $-(-c_j) = c_j$ and 0, are the increase to be made by the objective function if those variables were to become basic variables and assume a value equal to 1. Let us extend this premise a little further. For example, if x_1, which is now zero, enters the basis, the objective function increases by $-(-c_1) = c_1$ for each unit of x_1. Therefore, if $-c_1$ is negative, it is advisable to bring x_1 into this basis, since doing so results in the increase of z by c_1 for each unit of x_1 and the objective is to maximize z. However, if $-c_2$ is positive, it is more advantageous not to introduce x_2 into the basis. Therefore, the signs of the numbers in the last row provide the necessary means for an optimality test.

3. An optimality test. A simple sufficient condition for the current solution to be an optimal one is that the coefficients of the last row be nonnegative. If all the coefficients are nonnegative, the current solution is an optimal one. If on the other hand one of the coefficients is negative, the solution is not optimal. In this case, it is necessary to improve the solution by forcing the nonbasic variable associated with the negative value to become a basic variable. (An exception to this condition is given in Problem 14 at the end of this chapter.)

4. An improvement rule. If the current solution is not an optimal one, the simplex method provides a means for improving it. The improvement process consists in

1. Determining the variable that enters the basis,
2. Determining the variable that leaves the basis,
3. Updating the simplex tableau to make it canonical with respect to the new list of basic variables.

Note that any variable with a negative coefficient in the last row can enter the basis and improve the solution. Moreover, if there is more than one candidate to enter the basis, the variable with the largest contribution per unit is chosen. This can be determined by simply taking the variable with the most negative coefficient in the last row. Let x_s be the variable that is chosen to enter; then $-c_s = \text{minimum}\{-c_j\}$ and of course $-c_s < 0$.

Once it is decided which variable should enter ther basis, we must then determine the variable x_r that leaves the basis. In this instance, let a_{is} be the elements in the column of the variable that enters the basis. The variable that leaves the basis is determined by the smallest ratio of b_i over a_{is}, but only for positive a_{is}. Thus the example may be expressed as

$$\frac{b_r}{a_{rs}} = \underset{\substack{a_{is} > 0 \\ i = 1, 2, \ldots, m}}{\text{minimum}} \left\{ \frac{b_i}{a_{is}} \right\}$$

The reason for considering only $-c_i < 0$ is to bring profit to the objective function, and $-c_i > 0$ decreases the objective function. Similarly, in Step 2 only positive denominators are considered in order to maintain $b_i \geq 0$ throughout the algorithm.

Step 3 in the process is to perform the transformation of replacing x_r by x_s as a basic variable. This is easily accomplished by bringing the simplex tableau to a canonical form with respect to the new set of basic variables. The operation itself is called *pivoting* and the pivot element is a_{rs}.

The rules for the pivot operation can be summarized as:

1. Divide the rth row by a_{rs}.
2. Replace the sth column by zeros (except for the rth row, which should be one).
3. Compute any other element in the new table as follows:

$$\text{new } a_{ij} = \text{former } a_{ij} - \frac{a_{is} \cdot a_{rj}}{a_{rs}}$$

The c_b column remains the same except for a change of c_r by c_s.

The computational procedure for the simplex algorithm is summarized in Figure 3.1.

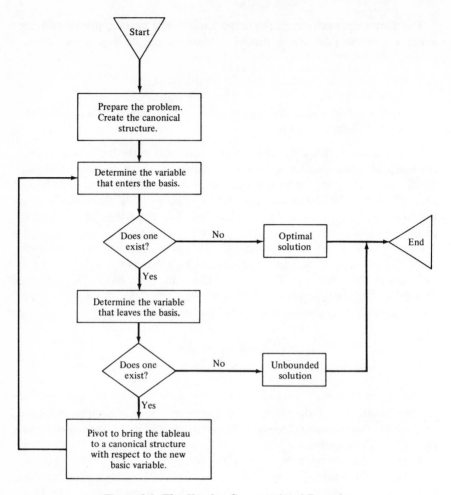

Figure 3.1. The Simplex Computational Procedure

3.3. A Maximization Example

We now use the simplex method to solve the Exclusive Furniture Company problem of Section 2.3. The problem is

$$
\begin{aligned}
\text{Maximize} \qquad & z = 5x + 7y \\
\text{subject to} \qquad & 2x + 5y \leqslant 50 \\
& 3x + 2y \leqslant 42 \\
& x \geqslant 0 \qquad y \geqslant 0
\end{aligned}
\tag{3.3}
$$

The first step in the process is to convert the constraints of (3.3) to equalities. After adding slack variables the problem is presented as

$$\text{Maximize} \qquad z = 5x + 7y + 0s_1 + 0s_2$$

$$\text{subject to} \qquad 2x + 5y + s_1 + 0s_2 = 50$$

$$3x + 2y + 0s_1 + s_2 = 42$$

$$x \geq 0 \qquad y \geq 0 \qquad s_1 \geq 0 \qquad s_2 \geq 0$$

An initial starting solution is

$$x = y = 0 \qquad s_1 = 50 \qquad s_2 = 42 \qquad z = 0$$

Note that this solution corresponds to the extreme point $(0,0)$ of Figure 2.6.

The initial simplex tableau appears in Table 3.2. The basic variables are $s_1 = 50$, $s_2 = 42$; the nonbasic variables are $x = 0$, $y = 0$.

The elements of the last row are calculated as follows:

$$0 \cdot 2 + 0 \cdot 3 - 5 = -5$$
$$0 \cdot 5 + 0 \cdot 2 - 7 = -7$$
$$0 \cdot 1 + 0 \cdot 0 - 0 = 0$$
$$0 \cdot 0 + 0 \cdot 1 - 0 = 0$$

The value of the objective function is calculated as follows:

$$0 \cdot 50 + 0 \cdot 42 = 0$$

From inspection of the initial tableau, it is evident that the solution is not optimal because there are negative values in the last row. Both coefficients of x and y are negative, and the more negative one is minimum$\{-5, -7\} = -7$. Thus, y enters the basis.

Table 3.2 Exclusive Furniture Company Problem Simplex Tableau

c_b	Basic variables	c_j Value	5 x	7 y	0 s_1	0 s_2
0	s_1	50	2	⑤	1	0
0	s_2	42	3	2	0	1
		0	−5	−7	0	0
7	y	10	2/5	1	1/5	0
0	s_2	22	⑪/5	0	−2/5	1
		70	−11/5	0	7/5	0
7	y	6	0	1	3/11	−2/11
5	x	10	1	0	−2/11	5/11
		92	0	0	1	1

The variable that leaves the basis is determined by taking the minimum ratio of b_i/a_{is}:

$$\text{Minimum}\left\{\frac{50}{5},\frac{42}{2}\right\}=\frac{50}{5}=\frac{b_1}{a_{12}}$$

In short, the variable that leaves the basis is s_1 and the pivot element is a_{12}, which is circled in Table 3.2.

The pivot operation is accomplished by dividing the first row by the pivot element $a_{12}=5$ and putting zero in the rest of the y column. All other elements are computed as follows:

$$b_2=42-\frac{50\cdot2}{5}=22$$

$$z=0-\frac{50\cdot(-7)}{5}=70$$

$$a_{21}=3-\frac{2\cdot2}{5}=\frac{11}{5}$$

$$a_{23}=0-\frac{2\cdot1}{5}=-\frac{2}{5}$$

$$a_{24}=1-\frac{2\cdot0}{5}=1$$

$$\text{Coefficient of } x \text{ in the last row}=-5-\frac{-7\cdot2}{5}=-\frac{11}{5}$$

$$\text{Coefficient of } s_1 \text{ in the last row}=0-\frac{-7\cdot1}{5}=\frac{7}{5}$$

$$\text{Coefficient of } s_2 \text{ in the last row}=0-\frac{-7\cdot0}{5}=0$$

At the end of the first iteration the solution is

$$y=10 \qquad s_2=22$$

as basic variables and

$$x=0 \qquad s_1=0$$

as nonbasic variables. Furthermore, the value of the objective function has been increased from 0 to 70. This new solution, $x=0$, $y=10$, is an extreme point in Figure 2.6. The new extreme point is adjacent to the previous extreme point of the starting solution. Therefore, the first iteration of the simplex procedure resulted in moving from one extreme point $(0,0)$ to an adjacent extreme point $(0,10)$. In general, this is a typical occurrence and by performing this process the solution is improved by 70 units.

The end of the first iteration does not signify the end of the procedure since there is still a negative value in the last row. In this instance, there is only one negative value in the last row. Thus, x is the entering variable. In addition, the minimum ratio determines the name of the variable that leaves the basis.

$$\text{Minimum}\left\{\frac{10}{2/5}, \frac{22}{11/5}\right\} = \frac{22}{11/5} = 10$$

The minimum ratio occurs in the second row. Therefore s_2 is the variable that leaves the basis. The transformation process involves the row replacement of variables s_2 by x. In this case, the pivot element is $a_{21} = 11/5$, and is circled for clear identification. In order to complete the transformation, the second row is now divided by the pivot element $a_{21} = 11/5$, and zeros are inserted in the first column except in row 2. The table that is generated from the procedure is as follows:

$$b_1 = 10 - \frac{22 \cdot (2/5)}{11/5} = 6$$

$$z = 70 - \frac{22 \cdot (-11/5)}{11/5} = 92$$

$$a_{12} = 1 - \frac{(2/5) \cdot 0}{11/5} = 1$$

$$a_{13} = \frac{1}{5} - \frac{2/5(-2/5)}{11/5} = \frac{3}{11}$$

$$a_{14} = 0 - \frac{(2/5) \cdot 1}{11/5} = -\frac{2}{11}$$

$$\text{Coefficient of } y = 0 - \frac{(-11/5) \cdot 0}{11/5} = 0$$

$$\text{Coefficient of } s_1 = \frac{7}{5} - \frac{(-11/5)(-2/5)}{11/5} = 1$$

$$\text{Coefficient of } s_2 = 0 - \frac{(-11/5) \cdot 1}{11/5} = 1$$

Therefore, the current basic variables are $x = 10$, $y = 6$, and the value of the objective function has increased from 70 to 92. In addition, the nonbasic variables are $s_1 = 0$, $s_2 = 0$.

It may be further noted that the current solution $(10, 6)$ is an extreme point in Figure 2.6. Once again, the simplex algorithm resulted in transiting from one extreme point $(0, 10)$ to an adjacent extreme point $(10, 6)$. The current solution, $x = 10$, $y = 6$ is also optimal since the coefficients in the last row are nonnegative.

3.4. A Minimization Example

Even though in the preceding examples the basics of the simplex method have been applied to a maximization problem, there exists a variation of the simplex for a minimization problem. Therefore, the next numerical example serves several purposes. First, it presents a more complicated example than the two-constraint, two-variable case in Section 3.3. Also, it presents a minimization problem and suggests two different methods for solving it.

The linear programming problem is

$$\text{Minimize} \qquad z = 4x_1 - x_2 - x_3 + 2x_4 + x_5$$

$$\text{subject to} \qquad
\begin{aligned}
x_1 + x_2 + x_3 + x_4 + x_5 &\leqslant 15 \\
-x_1 + 2x_2 - x_3 + 2x_4 &\leqslant 10 \qquad (3.4)\\
2x_1 + x_2 - x_3 + x_4 - 2x_5 &\leqslant 4
\end{aligned}$$

$$x_1, x_2, x_3, x_4, x_5 \geqslant 0$$

There are two avenues to a solution. The first is to convert the minimization problem to a maximization problem, and the second is to develop a new set of rules for a minimization objective function.

Conversion of a minimization to a maximization problem. Suppose that we want to find the minimum of three numbers— say, minimum$\{5, -3, 7\} = -3$. It is also possible to obtain the desired result by employing the maximum of the negative of the values, maximum$\{-5, 3, -7\} = 3$. The only adjustment necessary is to reverse the sign of the final result. Thus, using the property that

$$\text{Min}\{x_1, x_2, \ldots, x_n\} = -\text{Max}\{-x_1, -x_2, \ldots, -x_n\},$$

we can then consider the relationship as a maximization of profit or a minimization of expenses for a corporation.

The implication of this statement is that instead of solving for the objective function of Problem (3.4) as minimization:

$$z = 4x_1 - x_2 - x_3 + 2x_4 + x_5$$

we can solve it as the maximization of

$$z = -4x_1 + x_2 + x_3 - 2x_4 - x_5$$

with an adjustment of signs upon completion of the problem. Obviously, the set of constraints remains consistent for both objective functions.

Problem (3.4) is now transformed by using the following new objective function.

Maximize $\qquad -4x_1+x_2+x_3-2x_4-x_5$

subject to $\qquad x_1 +x_2+x_3 +x_4 +x_5 \leqslant 15$

$$-x_1+2x_2-x_3+2x_4 \qquad \leqslant 10 \qquad (3.5)$$

$$2x_1 +x_2-x_3 +x_4-2x_5 \leqslant 4$$

$$x_1, x_2, x_3, x_4, x_5 \geqslant 0$$

The first step in the process is to add slack variables to the smaller side of the inequalities to obtain Equations (3.6).

Maximize $\qquad -4x_1+x_2+x_3-2x_4-x_5+0s_1+0s_2+0s_3$

subject to $\qquad x_1 +x_2 +x_3 +x_4 +x_5+s_1 \qquad\qquad = 15$

$$-x_1 +2x_2 -x_3 +2x_4 \qquad\qquad +s_2 \qquad = 10 \qquad (3.6)$$

$$2x_1 +x_2 -x_3 +x_4 -2x_5 \qquad\qquad +s_3=4$$

$$x_1, x_2, x_3, x_4, x_5 \geqslant 0 \qquad s_1, s_2, s_3 \geqslant 0$$

The initial solution for the problem is the following:

$$s_1 = 15 \qquad s_2 = 10 \qquad s_3 = 4 \quad \text{and} \quad z=0$$

$$x_1 = x_2 = x_3 = x_4 = x_5 = 0$$

The data are then arranged in the initial simplex tableau shown in Table 3.3. In addition, the coefficients of the variables in the last row are calculated as follows:

$$x_1: 0 \cdot 1 + 0(-1) + 0 \cdot 2 - (-4) = 4$$

$$x_2: 0 \cdot 1 + 0 \cdot 2 + 0 \cdot 1 - 1 = -1$$

$$x_3: 0 \cdot 1 + 0(-1) + 0(-1) - 1 = -1$$

$$x_4: 0 \cdot 1 + 0 \cdot 2 + 0 \cdot 1 - (-2) = 2$$

$$x_5: 0 \cdot 1 + 0 \cdot 0 + 0(-2) - (-1) = 1$$

$$z: 0 \cdot 15 + 0 \cdot 10 + 0 \cdot 4 = 0$$

From this example it is apparent that the initial tableau is not optimal, since two of the coefficients are negative. In order to determine the variable that enters the basis, it is necessary to take the most negative coefficient of the last row, which is

$$\text{Minimum}\{-1, -1\} = -1$$

The coefficients of both x_2 and x_3 are -1; therefore, a decision is arbitrarily made to let x_2 enter the basis. A minimum ratio must now be taken

Table 3.3 Minimization Example—Simplex Tableau

c_b	Basic variables	c_j Value	-4 x_1	1 x_2	1 x_3	-2 x_4	-1 x_5	0 s_1	0 s_2	0 s_3	
0	s_1	15	1	1	1	1	1	1	0	0	
0	s_2	10	-1	2	-1	2	0	0	1	0	Initial
0	s_3	4	2	①	-1	1	-2	0	0	1	tableau
		0	4	-1	-1	2	1	0	0	0	
0	s_1	11	-1	0	2	0	3	1	0	-1	
0	s_2	2	-5	0	①	0	4	0	1	-2	Cycle 1
1	x_2	4	2	1	-1	1	-2	0	0	1	
		4	6	0	-2	3	-1	0	0	1	
0	s_1	7	⑨	0	0	0	-5	1	-2	3	
1	x_3	2	-5	0	1	0	4	0	1	-2	Cycle 2
1	x_2	6	-3	1	0	1	2	0	1	-1	
		8	-4	0	0	3	7	0	2	-3	
-4	x_1	7/9	1	0	0	0	$-5/9$	1/9	$-2/9$	③/⑨	
1	x_3	53/9	0	0	1	0	11/9	5/9	$-1/9$	$-3/9$	Cycle 3
1	x_2	75/9	0	1	0	1	3/9	3/9	3/9	0	
		100/9	0	0	0	3	43/9	4/9	10/9	$-15/9$	
0	s_3	7/3	3	0	0	0	$-5/3$	1/3	$-2/3$	1	Cycle 4
1	x_3	20/3	1	0	1	0	2/3	2/3	$-1/3$	0	Final
1	x_2	25/3	0	1	0	1	1/3	1/3	1/3	0	tableau
		15	5	0	0	3	2	1	0	0	

between b_i and a_{i2} in order to determine the variable that leaves the basis.

$$\text{Minimum}\left\{\frac{15}{1}, \frac{10}{2}, \frac{4}{1}\right\} = \frac{4}{1}$$

Thus, s_3 leaves the basis and is replaced by x_2. The tableau is now revised according to the previous rules of Section 3.2, and a new simplex tableau for cycle 1 is presented.

It is evident that the second tableau is not optimal. The variable that enters the basis is x_3, since

$$\text{Minimum}\{-2, -1\} = -2$$

and since using the column x_3 we find that

$$\text{Minimum}\left\{\frac{11}{2}, \frac{2}{1}\right\} = \frac{2}{1}$$

hence the variable s_2 leaves the basis. It is important to point out that the minimum ratio is taken for only the first two rows because $a_{33} = -1$ is negative, and should not be considered.

In this case, the value of the objective function increases from $z=0$ in the initial tableau to $z=4$ at the conclusion of the second tableau to $z=8$ at the completion of the third tableau. For the next iteration, z proceeds to $100/9$ and ultimately to 15. At cycle 3, variable x_1 replaces variable s_1 and immediately following is cycle 4 with variable s_3 replacing variable x_1. This is an example of a case where a variable (x_1) that enters the basis leaves it in the next iteration. The final basic variables are

$$x_2 = \frac{25}{3} \qquad x_3 = \frac{20}{3} \qquad s_3 = \frac{7}{3} \quad \text{and} \quad z = 15$$

The nonbasic variables are

$$x_1 = x_4 = x_5 = s_1 = s_2 = 0$$

Upon completion of cycle 4 it is evident that the solution is optimal, since all coefficients in the last row are nonnegative.

It now remains to reverse the sign of the objective function, so the value is $z = -15$. For a further check, consider

$$z = \text{Minimize } 4x_1 - x_2 - x_3 + 2x_4 + x_5$$

$$= 4 \cdot 0 - \frac{25}{3} - \frac{20}{3} + 2 \cdot 0 + 0 = -15$$

Simplex rules for minimization problems. The method presented in the preceding example, converting a minimization problem into a maximization problem, is a legitimate method. Furthermore, it is an easy one to use for anyone not previously exposed to the concepts of linear programming. There is also a second procedure for solving the minimization problem. In fact, when the problem is formulated as minimization, the only difference from a maximization problem is the optimality criterion. The optimality criterion for a maximization is nonnegativity in the last row, whereas the optimality criterion for a minimization problem is nonpositivity in the last row. Consequently, the variable that enters the basis at each iteration is the variable with the most positive value in the last row.

The next numerical example demonstrates the solution of a minimization problem with a new optimality criterion.

$$\begin{aligned} \text{Minimize} \quad & z = -3x_1 + 8x_2 - 5x_3 \\ \text{subject to} \quad & -x_1 \qquad\quad -2x_3 \leqslant 5 \\ & 2x_1 - 3x_2 + x_3 \leqslant 3 \\ & 2x_1 - 5x_2 + 6x_3 \leqslant 5 \\ & x_1, x_2, x_3 \geqslant 0 \end{aligned}$$

After reviewing this example, an astute observer might conclude that the first constraint is not important. For nonnegative values of x_1 and x_3, the first constraint is always satisfied. We can then proceed to solve the problem, ignoring the first constraint, and verify that the optimal solution satisfies the first constraint. However, suppose we overlook the fact that the first constraint can be omitted and include it as part of the problem. The first step is to transform the constraints to equalities.

Minimize $\quad z = -3x_1 + 8x_2 - 5x_3 + 0s_1 + 0s_2 + 0s_3$

subject to

$$-x_1 + 0x_2 - 2x_3 + s_1 \qquad = 5$$

$$2x_1 - 3x_2 + x_3 \quad + s_2 \quad = 3$$

$$2x_1 - 5x_2 + 6x_3 \qquad + s_3 = 5$$

$$x_1, x_2, x_3, s_1, s_2, s_3 \geqslant 0$$

The initial simplex tableau, as well as the solution, is illustrated in Table 3.4. The last row is then generated:

$$x_1: 0 \cdot (-1) + 0 \cdot 2 + 0 \cdot 2 - (-3) = 3$$

$$x_2: 0 \cdot 0 + 0 \cdot (-3) + 0 \cdot (-5) - 8 = -8$$

$$x_3: 0 \cdot (-2) + 0 \cdot 1 + 0 \cdot 6 - (-5) = 5$$

$$z: 0 \cdot 5 + 0 \cdot 3 + 0 \cdot 5 = 0$$

Obviously, the initial tableau is not optimal, since there are two positive elements in the last row. Furthermore, the name of the variable to enter the basis is determined by maximum $\{3,5\} = 5$, which is the most positive coefficient in the last row and belongs to x_3.

The variable to leave the initial solution is found by the ratio test:

$$\text{Minimum}\left\{\frac{3}{1}, \frac{5}{6}\right\} = \frac{5}{6}$$

which yields s_3. Recall, however, that because $a_{13} = -2$ is less than zero, it is

Table 3.4 Minimization Example—New Entering Variable Criterion

c_b	Basic variables	c_j Value	-3 x_1	8 x_2	-5 x_3	0 s_1	0 s_2	0 s_3	
0	s_1	5	-1	0	-2	1	0	0	
0	s_2	3	2	-3	1	0	1	0	
0	s_3	5	2	-5	⑥	0	0	1	
		0	3	-8	5	0	0	0	
0	s_1	20/3	$-1/3$	$-5/3$	0	1	0	1/3	
0	s_2	13/6	⑤/3	$-13/6$	0	0	1	$-1/6$	
-5	x_3	5/6	1/3	$-5/6$	1	0	0	1/6	
		$-25/6$	4/3	$-23/6$	0	0	0	$-5/6$	
0	s_1	71/10	0	$-21/10$	0	1	1/5	3/10	
-3	x_1	13/10	1	$-13/10$	0	0	3/5	$-1/10$	Optimal
-5	x_3	2/5	0	$-2/5$	1	0	$-1/5$	1/5	
		$-59/10$	0	$-21/10$	0	0	$-4/5$	$-7/10$	

not considered for the minimum ratio. The pivot element is $a_{33} = 6$, for replacing s_3 by x_3. The end of the first cycle, however, results in a tableau that is not optimal since x_1 has a positive coefficient in the last row. The value x_1 then enters the basis and s_2 leaves, since

$$\text{Minimum}\left\{ \frac{13/6}{5/3}, \frac{5/6}{1/3} \right\} = \frac{13/6}{5/3}$$

The pivot element is $a_{21} = 5/3$, and s_2 is replaced by x_1. The third and final tableau is presented, and the optimal solution is

$$s_1 = \frac{71}{10} \qquad x_1 = \frac{13}{10} \qquad x_3 = \frac{2}{5} \qquad z = -\frac{59}{10}$$

In summary, the optimality criterion is satisfied since all elements in the last row are nonpositive.

3.5. Surplus and Artificial Variables

So far, the constraints indicated in the examples have been in the form of "less than or equal to" ($\leq$). The initial simplex algorithm requires the conversion of the inequalities into equalities. An easy way to accomplish this task is by adding slack variables, as was done earlier. The slack variables have two purposes: The first is the conversion of inequalities to equalities; the second is to provide an initial feasible starting solution. If the constraint is of the form

$$a_{i1}x_1 + a_{i2}x_2 + \cdots + a_{in}x_n \leq b_i$$

then adding the slack variable results in

$$a_{i1}x_1 + a_{i2}x_2 + \cdots + a_{in}x_n + s_i = b_i$$

Therefore, an easy starting solution is $x_1 = x_2 = \cdots = x_n = 0$ and $s_i = b_i$, and provided that $b_i \geq 0$ and the $s_i \geq 0$, the solution is feasible.

In many instances, linear programming problems include constraints that are "greater than or equal to" ($\geq$). For this situation, it is necessary to *subtract* the nonnegative surplus $s_i \geq 0$ from the left-hand side of the inequality (the larger side). For example,

$$a_{i1}x_1 + a_{i2}x_2 + \cdots + a_{in}x_n \geq b_i$$

changes to

$$a_{i1}x_1 + a_{i2}x_2 + \cdots + a_{in}x_n - s_i = b_i$$

This change converts an inequality into an equality and fulfills the requirement of the simplex algorithm. However, there are difficulties, since there is no longer an easy feasible solution. The easy solution $x_1 = x_2 = \cdots = x_n = 0$ and $-s_i = b_i$ is no longer feasible because if b_i is strictly positive, s_i is strictly negative; but s_i is required to be nonnegative. In order to generate a feasible and easy starting solution, it is essential to introduce new variables, called *artificial variables*, $A_i \geqslant 0$. The sole purpose of artificial variables is to provide an easy feasible solution. Adding an artificial variable to an equation causes the equality to become unbalanced:

$$a_{i1}x_1 + a_{i2}x_2 + \cdots + a_{in}x_n - s_i = b_i$$

becomes

$$a_{i1}x_1 + a_{i2}x_2 + \cdots + a_{in}x_n - s_i + A_i = b_i$$

The addition of A_i to the left-hand side of the equality disrupts the equality. The only instance in which the equality holds is if $A_i = 0$, since the addition of a zero does not affect the equation. Therefore, the objective function is used to provide the means to push A_i down to zero. If the original objective function is

$$\text{Maximize} \quad z = c_1x_1 + c_2x_2 + \cdots + c_nx_n$$

it is necessary to attach a coefficient $-M$ to each artificial variable A_i that is added to the objective function. The $-M$ is a large negative number such as -10^6. The effect of this coefficient in a maximization problem is that each time A_i is positive, the objective function is "penalized" by $-M$, and the value of the objective function decreases drastically. It is because of the coefficient $-M$ that the simplex procedure drives A_i down to zero to avoid the high penalties and a decreased value of the objective function.

The idea is explained by the following example: The Pentex gas company mixes gasoline from Texas and Pennsylvania. Each gallon of Texas oil contains 30% high test and 70% regular, while each gallon of Pennsylvania oil contains 50% high test, 30% regular, and 20% heating oil. The company has a contract to supply at least 2000 gallons of high test and 3000 gallons of regular gasoline. One gallon of Texas oil costs 7¢ and one gallon of Pennsylvania oil costs 9¢. How many gallons of Texas and Pennsylvania oil should Pentex purchase to meet its obligations? Assume that x gallons of Texas and y gallons of Pennsylvania are purchased. The cost is then $7x + 9y$. The high test gas produced is $0.3x + 0.5y$.

$$0.3x + 0.5y \geqslant 2000 \text{ gallons}$$

The regular gas produced is $0.7x + 0.3y$.

$$0.7x + 0.3y \geqslant 3000$$
$$x \geqslant 0 \quad y \geqslant 0$$

(The heating oil provides no contribution.) The mathematical formulation is now as follows:

$$\begin{aligned}
\text{Minimize} \quad & 7x + 9y \\
\text{subject to} \quad & 0.3x + 0.5y \geqslant 2000 \\
& 0.7x + 0.3y \geqslant 3000 \\
& x, y \geqslant 0
\end{aligned} \tag{3.7}$$

The first step is to convert the inequalities into equalities by subtracting the surplus variables s_i from the larger (left-hand) side of the inequalities. Also, the objective function is changed from a minimization to a maximization problem by changing the signs of its coefficients. Thus Problem (3.7) becomes (3.8).

$$\text{Maximize} \qquad -7x-9y+0s_1+0s_2$$

$$\text{subject to} \qquad 0.3x+0.5y-s_1 \qquad =2000 \qquad (3.8)$$

$$0.7x+0.3y \qquad -s_2 =3000$$

$$x,\,y,\,s_1,\,s_2 \geqslant 0$$

Since the surplus variables yield negative values, Problem (3.8) does not have an easy feasible starting solution. The solution $x=y=0$, and $-s_1=2000$, $-s_2=3000$ is infeasible. In order to provide an easy feasible solution it is necessary to introduce artificial variables, $A_1 \geqslant 0$, $A_2 \geqslant 0$. These variables destroy the balance of the equalities as long as A_i is positive. To eliminate the artificial variables from the solution, a penalty of $-M$ (in a maximization) is assigned to each A_i. The linear programming problem is now illustrated in (3.9).

$$\text{Maximize} \qquad z=-7x-9y+0s_1+0s_2-MA_1-MA_2$$

$$\text{subject to} \qquad 0.3x+0.5y-s_1 \quad +A_1 \quad =2000 \qquad (3.9)$$

$$0.7x+0.3y \quad -s_2 \quad +A_2=3000$$

$$x,\,y,\,s_1,\,s_2,\,A_1,\,A_2 \geqslant 0$$

The easy basic feasible starting solution is now $x=y=s_1=s_2=0$, $A_1=2000$, $A_2=3000$, and the objective function is $z=-5000M$ with M a large positive number.

The initial simplex tableau is presented in Table 3.5. The last row is calculated as before where M is considered a large number:

for x $\qquad -M\cdot0.3-M\cdot0.7-(-7)=-M+7$

for y $\qquad -M\cdot0.5-M\cdot0.3-(-9)=-0.8M+9$

The initial table is not optimal, and the most negative coefficient in the last row is $-M+7(-M+7<-0.8M+9)$, and x enters the basis. The variable to leave the basis is A_2 since

$$\text{Minimum} \left\{ \frac{2000}{0.3}, \frac{3000}{0.7} \right\} = \frac{3000}{0.7}$$

At the end of the first cycle, the solution increases from $-5000M$ to $-(5000M/7)-30,000$, but it is still not optimal. The variable s_2 replaces A_1 at the second cycle, and y replaces s_2 at the third cycle. The third cycle produces an optimal solution:

$$x=\frac{45000}{13}, \qquad y=\frac{25000}{13} \qquad s_1=s_2=A_1=A_2=0$$

Table 3.5 Pentex Gas Problem—Simplex Tableau

c_b	Basic variables	c_j Value	-7 x	-9 y	0 s_1	0 s_2	$-M$ A_1	$-M$ A_2	
$-M$	A_1	2000	0.3	0.5	-1	0	1	0	
$-M$	A_2	3000	(0.7)	0.3	0	-1	0	1	
		$-5000M$	$-M+7$	$-0.8M+9$	M	M	0	0	
$-M$	A_1	$\dfrac{5000}{7}$	0	$\dfrac{26}{70}$	-1	$\left(\dfrac{3}{7}\right)$	1	$\dfrac{-3}{7}$	
-7	x	$\dfrac{30,000}{7}$	1	$\dfrac{3}{7}$	0	$\dfrac{-10}{7}$	0	$\dfrac{10}{7}$	Cycle 1
		$\dfrac{-5000M}{7}-30,000$	0	$-\dfrac{26}{70}M+6$	M	$-\dfrac{3}{7}M+10$	0	$\dfrac{10}{7}M-10$	
0	s_2	$\dfrac{5000}{3}$	0	$\left(\dfrac{26}{30}\right)$	$\dfrac{-7}{3}$	1	$\dfrac{7}{3}$	-1	
-7	x	$\dfrac{20,000}{3}$	1	$\dfrac{5}{3}$	$\dfrac{-10}{3}$	0	$\dfrac{10}{3}$	0	Cycle 2
		$\dfrac{-140,000}{3}$	0	$\dfrac{-8}{3}$	$\dfrac{70}{3}$	0	$M-\dfrac{70}{3}$	M	
-9	y	$\dfrac{25,000}{13}$	0	1	$\dfrac{-35}{13}$	$\dfrac{15}{13}$	$\dfrac{35}{13}$	$\dfrac{-15}{13}$	
-7	x	$\dfrac{45,000}{13}$	1	0	$\dfrac{15}{13}$	$\dfrac{-25}{13}$	$\dfrac{-15}{13}$	$\dfrac{25}{13}$	Cycle 3
		$\dfrac{-540,000}{13}$	0	0	$\dfrac{210}{13}$	$\dfrac{40}{13}$	$M-\dfrac{210}{13}$	$M-\dfrac{40}{13}$	

and the objective function $z=-540,000/13$. It is important to remember that Problem (3.7) is a minimization problem, and therefore the value of the original objective function is $540,000/13$. As anticipated, the artificial variables A_1 and A_2 have diminished to zero after serving the purpose of providing an initial basic feasible solution. The high penalties of $-M$ eliminated the artificial variables from the optimal solution.

As illustrated in the foregoing example, each time there is no starting basic feasible solution, an artificial variable is added to at least one constraint. In some instances the constraints have slack variables that appear in the basic solution, while in other instances there are no slack variables that are basic. In the latter cases, artificial variables are added to the constraints.

Suppose, for example, that the problem is

$$\text{Maximize} \qquad -7x-9y$$
$$\text{subject to} \qquad 0.3x+0.5y \geqslant 2000$$
$$0.7x+0.3y \geqslant 3000$$
$$x+2y \leqslant 50,000$$
$$0.2x+0.1y \leqslant 6000$$
$$x, y \geqslant 0$$

After conversion of inequalities to equalities, two artificial variables are necessary and the problem is stated as:

$$\text{Maximize} \qquad -7x - 9y + 0s_1 + 0s_2 + 0s_3 + 0s_4 - MA_1 - MA_2$$

$$\text{subject to} \qquad 0.3x + 0.5y - s_1 + A_1 \qquad\qquad = 2000$$

$$0.7x + 0.3y \qquad -s_2 + A_2 \qquad = 3000$$

$$x + 2y \qquad\qquad +s_3 \qquad = 50000$$

$$0.2x + 0.1y \qquad\qquad\qquad +s_4 = 6000$$

$$x, y, s_1, s_2, s_3, s_4, A_1, A_2 \geqslant 0$$

The basic artificial starting solution is

$$x = y = s_1 = s_2 = 0 \qquad A_1 = 2000 \qquad A_2 = 3000 \qquad s_3 = 50{,}000 \qquad s_4 = 6000$$

3.6. Special Cases: Degeneracy, Alternative, Unbounded, and Infeasible Solutions

So far, the numerical examples have been structured in such a way that an optimal solution always exists. In this section, unique situations in linear programming are discussed in detail. It is useful to recognize these special cases and to be able to identify the symptoms of each specific case. The four unique cases are degenerate, alternate, unbounded, and infeasible solutions to a linear programming problem.

3.6.1. Degeneracy

A basic solution to a linear programming problem with m constraints and n variables contains precisely m basic variables and $n - m$ nonbasic variables. All nonbasic variables are equal to zero. It is possible, however, that one or more of the m basic variables is equal to zero. If this is the case, then the solution is degenerate, which means that there are less than m positive variables in the solution. Degeneracy may occur at any iteration in the course of the simplex algorithm, including the final tableau.

Degeneracy in linear programming is traced to the determination of the variable to leave the basis: If in taking the minimum ratio of b_i / a_{is}, there is not a unique minimum but a tie between two or more variables, there will be a degenerate solution in the next iteration. A numerical example is the following.

$$\text{Maximize} \qquad 5x + 2y + 4z$$

$$\text{subject to} \qquad 2x + 0y + 3z \leqslant 8$$

$$4x + y + 8z \leqslant 16$$

$$x + 7y + 6z \leqslant 5$$

$$x, y, z \geqslant 0$$

After addition of the slack variables s_1, s_2, and s_3, the initial simplex tableau is presented in Table 3.6. The coefficient of x in the last row is the most negative, and therefore x enters the basis. To determine the variable that leaves the basis, a minimization of ratios is taken:

$$\text{Minimum}\left\{ \frac{8}{2}, \frac{16}{4}, \frac{5}{1} \right\} = 4$$

This minimum occurs in both the first and second rows, and there is a choice; x can replace s_1 or s_2 in the basis. In both cases, s_1 and s_2 decrease to the zero level. In this sense, the choice is whether to retain s_1 or s_2 at the zero level in the basis. However, the tie can be arbitrarily broken by replacing s_1 with x. The new simplex tableau with the basic variables x, s_2, s_3 is presented in Table 3.6. Note that $s_1 = y = z = 0$ are nonbasic variables, and $s_2 = 0$ is a basic variable. The simplex algorithm continues in the usual manner, and the next variable to enter the basis is y (-2 is the coefficient in the last row). In order to determine the variable that leaves use column y and find the minimum ratio.

$$\text{Minimum}\left\{ \frac{0}{1}, \frac{1}{7} \right\} = \frac{0}{1}$$

($4/0$ is eliminated because the denominator is not positive). Hence y replaces s_2, and in the next simplex tableau the solution is still degenerate

Table 3.6 A Degenerate Problem

c_b	Basic variables	c_j Value	5 x	2 y	4 z	0 s_1	0 s_2	0 s_3
0	s_1	8	②	0	3	1	0	0
0	s_2	16	4	1	8	0	1	0
0	s_3	5	1	7	6	0	0	1
		0	-5	-2	-4	0	0	0
5	x	4	1	0	$\frac{3}{2}$	$\frac{1}{2}$	0	0
0	s_2	0	0	①	2	-2	1	0
0	s_3	1	0	7	$4\frac{1}{2}$	$\frac{-1}{2}$	0	1
		20	0	-2	$3\frac{1}{2}$	$2\frac{1}{2}$	0	0
5	x	4	1	0	$\frac{3}{2}$	$\frac{1}{2}$	0	0
2	y	0	0	1	2	-2	1	0
0	s_3	1	0	0	$-9\frac{1}{2}$	⑬$\frac{1}{2}$	-7	1
		20	0	0	$7\frac{1}{2}$	$-1\frac{1}{2}$	2	0
5	x	107/27	1	0	50/27	0	7/27	$-1/27$
2	y	4/27	0	1	16/27	0	$-1/27$	4/27
0	s_1	2/27	0	0	$-19/27$	1	$-14/27$	2/27
		543/27	0	0	174/27	0	33/27	3/27

with $x=4$, $y=0$, $s_3=1$ as basic variables. After an additional iteration, an optimal solution is reached, and the solution is no longer degenerate. It is possible to have degenerate solutions on the way to optimality, and the optimal solution may or may not (as in the example) be degenerate.

A complication that can result from degenerate solutions is *cycling*, in which the same basis is repeated again and again, so that the process never terminates. If a degenerate solution does exist, and an arbitrary means is used to break ties, it is conceivable that the solution will repeat itself and cycling will occur. There are, however, systematic methods for breaking ties so that cycling does not occur. (See Charnes and Cooper, 1965, p. 416.)

For a linear programming problem with m constraints, the decision maker expects to have m positive variables. Thus, from the n possible products (decision variables), the optimal solution contains only m products. (Some of the basic variables may be slack or surplus variables.) Some "advantage factors" can be found in the degenerate solution when less than m products are positive. The decision maker is then in a position to concentrate on fewer products and receive the same optimal profit.

A geometric interpretation of degenerate solutions is given in the two-dimensional problem of Figure 3.2. The problem is to find x, y that

$$\begin{aligned}
\text{Maximize} \qquad & x+y \\
\text{subject to} \qquad & x+y \leqslant 2 \\
& x \quad\;\; \leqslant 1 \\
& y \leqslant 1 \\
& x, y \geqslant 0
\end{aligned}$$

The solution in Figure 3.2 is $x=y=1$, $z=2$. After addition of the slack variables the problem is

$$\begin{aligned}
\text{Maximize} \quad & x+y+0s_1+0s_2+0s_3 \\
\text{subject to} \quad & x+y+s_1 \qquad\quad =2 \\
& x \qquad +s_2 \quad =1 \\
& y \qquad\quad +s_3=1 \\
& x, y, s_1, s_2, s_3 \geqslant 0
\end{aligned}$$

The number of basic variables is three, and the number of nonbasic (zero) variables is two. But the optimal solution contains $x=y=1$ and $s_1=s_2=s_3=0$, two positive and three zero variables. The geometric reason is generally stated as: An optimal solution to a two-dimensional linear programming problem is the intersection of two constraints. But in Figure 3.2 the optimal solution is the intersection of three constraints. Moreover, the constraints all pass through the optimal solution $(1,1)$.

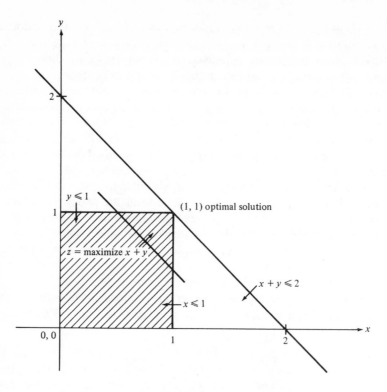

Figure 3.2. Graphical Example of Degeneracy

3.6.2. *Alternative Solutions*

For the decision maker, it is important to have alternative solutions to the one recommended. Let us not forget that the linear programming model is only a representation of a real-life problem. In many instances, it is impossible to consider all the constraints in the model. Quantitative restrictions are easy to express, but qualitative restrictions are difficult. Many factors—for example, a recession, a change in the taste of customers, a modification of the conservative approach of management—that exist in reality cannot be easily represented in the model. It is extremely helpful for the decision maker to be able to consider more than one solution. When several solutions are presented, the decision maker can choose the one that fulfills the requirements while satisfying other restrictions. (Some discussion of this subject was presented earlier in Section 2.5.)

The example to be considered here involves a car dealer who sells station wagons for a profit of $600 per car and sport cars for a profit of $1200 per car. The dealer has a parking lot of 10,000 ft^2. A station wagon requires

300 ft^2 for parking and a sport car requires only 100 ft^2. The price of a station wagon is \$5000 and a sport car is \$10,000. For insurance purposes, the total inventory of cars cannot exceed \$250,000. How many station wagons and sport cars should be sold each month to maximize the dealer's profit?

Let x be the number of station wagons and y the number of sport cars.

$$300x + 100y \leqslant 10,000 \quad \text{space constraint}$$

$$5000x + 10,000y \leqslant 250,000 \quad \text{insurance constraint}$$

$$\text{Maximize } z = 600x + 1200y \quad \text{profit}$$

$$x, y \geqslant 0$$

After addition of the slack variables s_1, s_2, the problem becomes

$$\text{Maximize} \qquad z = 600x + 1200y + 0s_1 + 0s_2$$

$$\text{subject to} \qquad 300x + 100y + s_1 \quad = 10,000$$

$$5000x + 10000y + s_2 = 250,000$$

$$x, y, s_1, s_2 \geqslant 0$$

The initial simplex tableau in Table 3.7 is not optimal, but after one iteration where y replaces s_2, an optimal solution is achieved. The optimal solution is to sell $y = 25$ sport cars, and eliminate the station wagons for a profit of \$30,000 per month.

Notice that at the end of the first iteration, the usual zeros are located at the basic variables in the last row. The basic variables are s_1 and y, and in fact there are zeros in these columns in the last row. But in addition, there is another zero in the last row under one of the nonbasic variables x. This is an indication that an alternative solution does exist.

Table 3.7 Car Dealer—Alternative Optimal Solutions

c_b	Basic variables	c_j Value	600 x	1200 y	0 s_1	0 s_2	
0	s_1	10,000	300	100	1	0	
0	s_2	250,000	5000	10,000	0	1	
		0	−600	−1200	0	0	
0	s_1	7500	250	0	1	−0.01	
1200	y	25	0.5	1	0	0.0001	Optimal
		30,000	0	0	0	0.12	
600	x	30	1	0	0.004	−0.00004	
1200	y	10	0	1	−0.002	0.00012	Optimal
		30,000	0	0	0	0.12	

Therefore, *alternative solutions* exist whenever the coefficient in the bottom row is a zero for one of the nonbasic variables.

The procedure used to generate this alternative solution is to force the variable x into the basis. Usually, if the coefficient of a variable is negative, its introduction to the basis effects an improvement in the objective function. However, since the coefficient is zero, there is no change in the value of the objective function as the variable enters the basis.

The procedure used to determine the variable that leaves the basis is the same as previously mentioned. Take the minimum ratio b_i/a_{is}:

$$\text{Minimum}\left\{\frac{7500}{250}, \frac{25}{0.5}\right\} = \frac{7500}{250} = 30$$

In this case s_1 leaves the basis, and a pivot operation around $a_{11} = 250$ will offer the second optimal solution of

$$x = 30 \qquad \text{station wagons}$$
$$y = 10 \qquad \text{sport cars}$$
$$z = \$30,000$$

It should be emphasized that the value of the objective function has not changed. The dealer can now decide whether to sell only sport cars or to mix both sport cars and station wagons. Hence, from the point of view of profit, the dealer has two equivalent choices, since the profit of $30,000 is the same for both cases.

3.6.3. Unbounded Solution

In most instances, real-life problems are of a bounded nature with a finite optimal solution. Usually, the reason for obtaining an unbounded solution is either inaccurate input data or a missing constraint. Therefore, it is important to be able to detect an unbounded solution, so that the input data and set of constraints can be verified for validity.

The symptoms associated with an unbounded solution are easy to recognize. A good example is when a column has a negative coefficient in the last row (which is an indication of improvement if that variable enters the basis), but none of the elements in that column are positive (so no variable can leave the basis). In short, an unbounded solution exists whenever there is a column with a negative coefficient in the last row and the column does not have a positive element.

From an intuitive perspective, it is evident that a variable wants to enter the basis in an effort to improve the objective function; however, none of the basic variables desires to leave. Consequently, the result would be an unbounded solution.

Table 3.8 An Unbounded Solution

c_b	Basic variables	c_j Value	2 x	3 y	1 z	0 s_1	0 s_2
0	s_1	5	1	2	-2	1	0
0	s_2	2	3	-1	-1	0	1
		0	-2	-3	-1	0	0

An example of this is presented next.

$$\text{Maximize} \qquad 2x+3y+z$$

$$\text{subject to} \qquad x+2y-2z \leqslant 5$$

$$3x-y-z \leqslant 2$$

$$x, y, z \geqslant 0$$

(Clearly, if $z=\infty$ the objective function is unbounded, and the constraints are satisfied.) After addition of the slack variables s_1, s_2, the problem may be written as:

$$\text{Maximize} \quad 2x+3y+z+0 \cdot s_1 + 0 \cdot s_2$$

$$\text{subject to} \quad x+2y-2z+s_1 \quad = 5$$

$$3x-y-z \quad +s_2 = 2$$

$$x, y, z, s_1, s_2 \geqslant 0$$

The initial simplex tableau is presented in Table 3.8. According to the rules of the simplex procedure, the most negative coefficient determines the variable to enter the basis. In this case, y enters the basis, since its coefficient is -3. In this problem, however, there is another variable z with a negative value, -1, in the last row and there are no positive elements in the column $(-2, -1)$. In this sense, the problem has an unbounded solution, as was anticipated before, with the variable z having a limitless value.

3.6.4. An Infeasible Solution

The last case in the group of irregular linear programming problems is when there is no feasible solution. In various circumstances, the constraints associated with problems may have been derived from different sources. It is then conceivable that the variations in the sources of information may result in conflicting restrictions on the system. A good instance of such conflict would be the case of one company vice-president demanding that the production level be set at 5000 units per month, and another vice-president restricting production to 3000 units per month. Obviously, it is impossible to satisfy both constraints simultaneously. In any event, when the constraints become more complicated it becomes more difficult to be aware of contradictions in the restrictions. From a different perspective, the production level may be limited by such hidden constraints as machine capability, availability of resources, or a limited financial budget. It is the intersection of the entire system's restrictions that determines whether or not a feasible solution exists.

A graphical representation of an infeasible solution was discussed in Section 2.5. Let us now extend this concept further with an algebraic representation of an infeasible solution.

If the problem under consideration has an easy feasible starting solution (i.e., the slack variables $s_i = b_i$, $s_i \geqslant 0$), then an optimal feasible solution will exist simply because the problem begins with a feasible solution.

However, a doubtful situation arises pertaining to the optimality of a solution when the starting solution includes artificial variables. In this instance, the artificial variables are added to an equality with the assumption that the variables will vanish to zero. However, if there is no way to remove the artificial variables from the basis, then the variables remain positive at the optimal solution, and an infeasible solution exists. The assumption based on the fact that all artificial variables will vanish at the zero level is erroneous. It is altogether possible that, even at the price of a high-penalty M in the objective function, the artificial variables will still remain positive in the basis. This occurs if there is no other available choice of variables to enter the basis and the artificial variables are then necessary (basic variables) for solving the equations.

Infeasible Solution. If the optimal solution contains positive artificial variables, a feasible solution does not exist. The concept described above is presented in the next example. The objective function is to

$$\text{Maximize} \qquad\qquad 3x + 2y$$

subject to the two constraints $x + 2y \leqslant 3$

$$x + y \geqslant 5$$

$$x, y \geqslant 0$$

After adding the slack variable to the first constraint and subtracting the surplus variable from the second constraint we have

$$\text{Maximize} \qquad 3x + 2y + 0s_1 + 0s_2$$

$$\text{subject to} \qquad x + 2y + s_1 \qquad = 3$$

$$x + y \qquad - s_2 = 5$$

$$x, y, s_1, s_2 \geqslant 0$$

It is evident that there is no easy starting solution, and therefore an artificial

Table 3.9 No Feasible Solution

c_b	Basic variables	c_j Value	3 x	2 y	0 s_1	0 s_2	$-M$ A_1	
0	s_1	3	①	2	1	0	0	
$-M$	A_1	5	1	1	0	-1	1	
		$-5M$	$-M-3$	$-M-2$	0	M	0	
3	x	3	1	2	1	0	0	
$-M$	A_1	2	0	-1	-1	-1	1	Optimal
		$-2M+9$	0	$M+4$	$M+3$	M	0	

variable must be introduced. Notice that only the second constraint requires an artificial variable because s_1 can serve as the starting solution for the first constraint. Thus, the problem becomes

$$\text{Maximize} \qquad z = 3x + 2y + 0s_1 + 0s_2 - MA_1$$

$$\text{subject to} \qquad x + 2y + s_1 \qquad\quad = 3$$

$$x + y \qquad -s_2 + A_1 = 5$$

$$x, y, s_1, s_2, A_1 \geqslant 0$$

The initial simplex tableau is presented in Table 3.9. The solution is not optimal, and x enters the basis since the coefficient in the last row is $-M-3 < 0$. The minimum ratio is $\{3/1, 5/1\} = 3$ and therefore s_1 leaves the basis. The simplex tableau at the end of the first iteration, is an optimal one. The optimal solution is $x = 3$, $A_1 = 2$, and the value of the objective function is $z = -2M + 9$. Since $A_1 > 0$, we now conclude that the problem is infeasible. Looking back at the constraints $x + 2y \leqslant 3$, $x + y \geqslant 5$, it is evident that a contradiction exists. Since $x + y \geqslant 5$, it is impossible for $x + y + y$ to be less than $3(y \geqslant 0)$.

We might wonder what the results in Table 3.9 would have been had y entered the basis instead of x. Needless to say, the solution is still infeasible. In other words, the order of the variables entering the basis is not going to change the final solution. After y enters the basis, x enters and replaces the y, resulting in the same infeasible solution.

3.7. Rules to Check for Algebraic Computational Errors

Real-life problems have a tendency to be large enough to justify the utilization of computers for finding solutions. A problem with more than 10 constraints and more than 30 variables is considered too large for linear programming using manual skills. Most computers, including a few mini-computers, have library programs that solve linear programming problems. The programs are written efficiently in an effort to save both computer time and storage space. In addition, most time-sharing systems have access to computers with linear programming facilities. Nevertheless, it is strongly recommended that you first acquire a basic understanding of linear programming at the manual level in order to master the algorithm. Those who have exposed themselves to the concepts are well aware of mathematical computational errors.

The purpose of this section is to provide a framework for double checking the simplex method. The following list of rules is associated with properties of the simplex algorithm. A violation of any of the rules would result from an algebraic mistake.

1. A variable that has left the basis cannot reenter in the next iteration. A variable must remain a nonbasic variable for at least one iteration before reentering the basis.
2. The value column must always contain nonnegative numbers (except in the objective function row). When you start with nonnegative numbers on the right-hand side, it is imperative that the value column continue to list nonnegative numbers throughout the procedure.
3. In a maximization problem, the value of the objective function must increase or remain the same from one simplex tableau to the next. (In a minimization problem, the value of the objective function must decrease or stay the same between any two consecutive iterations.)
4. At any iteration, the basic variables column must be the unit vector. That is, each column has exactly one 1 and the rest of the numbers are 0. For example, in Table 3.6 the final tableau contains the variables x, y, s_1. Therefore the columns under the variables x, y, s_1 are

$$\begin{pmatrix} 1 \\ 0 \\ 0 \\ 0 \end{pmatrix} \begin{pmatrix} 0 \\ 1 \\ 0 \\ 0 \end{pmatrix} \begin{pmatrix} 0 \\ 0 \\ 1 \\ 0 \end{pmatrix}$$

respectively.
5. The values of the last row can be calculated in two ways. One method is suggested at the end of Section 3.2 in the discussion of the pivot operation. For example, in the final tableau in Table 3.6 the coefficient in the last row for z is

$$7\tfrac{1}{2} - \frac{(-9\tfrac{1}{2}) \cdot (-1\tfrac{1}{2})}{13\tfrac{1}{2}} = \frac{174}{27}$$

and the coefficient for s_2 is

$$2 - \frac{(-7) \cdot (-1\tfrac{1}{2})}{13\tfrac{1}{2}} = \frac{33}{27}$$

The second method for calculating the coefficients is similar to the procedures used for evaluating the last row of the initial simplex tableau. The value of the coefficient is obtained by multiplying the cost row c_b by the column under any variable x_j, and then subtracting the original coefficient c_j. For example, in the last iteration the value for z is

$$5 \cdot \frac{50}{27} + 2 \cdot \frac{16}{27} + 0 \cdot \frac{-19}{27} - 4 = \frac{174}{27}$$

and for s_2 is

$$5 \cdot \frac{7}{27} + 2 \cdot \frac{-1}{27} + 0 \cdot \frac{-14}{27} - 0 = \frac{33}{27}$$

Both methods must present the same results.

(Rules 6 and 7 require knowledge of matrix inverses.)

6. The starting solution in the initial simplex tableau contains the basic variables s_1, s_2, s_3. At any iteration including the optimal one, under the same variables s_1, s_2, s_3 there is a matrix called B^{-1}. In the final tableau the matrix is

$$B^{-1} = \begin{bmatrix} 0 & 7/27 & -1/27 \\ 0 & -1/27 & 4/27 \\ 1 & -14/27 & 2/27 \end{bmatrix}$$

As shown in Rule 4, the final tableau contains the variables x, y, s_1. Among the same variables in the initial simplex tableau there is a matrix

$$B = \begin{bmatrix} 2 & 0 & 1 \\ 4 & 1 & 0 \\ 1 & 7 & 0 \end{bmatrix}$$

The multiplication of the two matrices B^{-1} and B will provide an identity matrix.

$$B^{-1}B = \begin{bmatrix} 0 & 7/27 & -1/27 \\ 0 & -1/27 & 4/27 \\ 1 & -14/27 & 2/27 \end{bmatrix} \begin{bmatrix} 2 & 0 & 1 \\ 4 & 1 & 0 \\ 1 & 7 & 0 \end{bmatrix} = \begin{bmatrix} 1 & 0 & 0 \\ 0 & 1 & 0 \\ 0 & 0 & 1 \end{bmatrix}$$

The multiplication of the two matrices must result in an identity. One matrix is found in the current tableau under the original basic variables, and the second matrix is found in the original tableau under the current basic variables.

7. The last rule provides the ability to calculate any column by the B^{-1} matrix, and the original vector. In this respect, the column for z

$$z = \begin{pmatrix} 50/27 \\ 16/27 \\ -19/27 \end{pmatrix}$$

can be calculated by multiplying the original column for z

$$z = \begin{pmatrix} 3 \\ 8 \\ 6 \end{pmatrix}$$

by B^{-1}, which yields the following:

$$\begin{bmatrix} 0 & 7/27 & -1/27 \\ 0 & -1/27 & 4/27 \\ 1 & -14/27 & 2/27 \end{bmatrix} \begin{pmatrix} 3 \\ 8 \\ 6 \end{pmatrix} = \begin{bmatrix} 50/27 \\ 16/27 \\ -19/27 \end{bmatrix}$$

The current column for s_2 can be verified by multiplying B^{-1} by the original column of s_2, $\begin{pmatrix} 0 \\ 1 \\ 0 \end{pmatrix}$:

$$\begin{bmatrix} 0 & 7/27 & -1/27 \\ 0 & -1/27 & 4/27 \\ 1 & -14/27 & 3/27 \end{bmatrix} \begin{pmatrix} 0 \\ 1 \\ 0 \end{pmatrix} = \begin{pmatrix} 7/27 \\ -1/27 \\ -14/27 \end{pmatrix}$$

which is the current column vector for s_2. The value column has the same property and we can verify the current value of that column by multiplying B^{-1} by the original value column, $\begin{pmatrix} 8 \\ 16 \\ 5 \end{pmatrix}$:

$$\begin{bmatrix} 0 & 7/27 & -1/27 \\ 0 & -1/27 & 4/27 \\ 1 & -14/27 & 2/27 \end{bmatrix} \begin{pmatrix} 8 \\ 16 \\ 5 \end{pmatrix} = \begin{pmatrix} 107/27 \\ 4/27 \\ 2/27 \end{pmatrix}$$

which is the current value column.

3.8. Summary

This chapter has presented the *simplex method*, which can be used to solve any linear programming problem. The size of the problems that can be solved is limited not by the method, as with the graphical solution, but rather, by computer hardware and software capabilities. The method is very straightforward once a *starting solution* has been found. In general, a search is made by moving from *basic solution* (extreme point) to an *adjacent solution*, either stopping because the *optimality test* has been passed or continuing to the next adjacent basic solution by using the *improvement* rule.

The starting solution is found by the use of additional variables that either take up the *slack* or the *surplus* or are purely *artificial* and necessary only temporarily. The simplex method rules clearly allow for the possibilities enumerated in the last chapter. That is, the method identifies problems with *no feasible* solutions, problems with *unbounded* solutions, and problems with *multiple* solutions.

In a sense, our exposition of linear programming is now complete. However, throughout the last two chapters it has been assumed that all of the coefficients in the problem are known. It may be that for some of the coefficients only ranges are known rather than exact figures. For these cases it is useful to have a more general solution than the one final tableau. The means to accomplish this is presented in the next chapter. In fact, the next chapter contains a complete analysis of methods for restarting the problem after a solution has been found.

References and Selected Readings

Charnes, A., and W. W. Cooper 1965. *Management Models and Industrial Applications of Linear Programming*. New York: John Wiley and Sons, Inc.

Cooper, L., and D. Steinberg 1974. *Methods and Applications of Linear Programming*. Philadelphia: Saunders.

Dantzig, G. B. 1963. *Linear Programming and Extensions*. Princeton, New Jersey: Princeton University Press.

Gass, S. I. 1975. *Linear Programming Methods and Applications*, 4th ed. New York: McGraw-Hill.

Hughes, A. J., and D. E. Grawiog 1973. *Linear Programming: An Emphasis on Decision Making*. Reading, Massachusetts: Addison-Wesley.

Kim, C. 1971. *Introduction to Linear Programming*. New York: Holt, Rinehart and Winston.

Levin, R. I., and P. Lamone Rudolph 1969. *Linear Programming for Management Decisions*. Homewood, Illinois: Irwin.

Sasaki, K. 1970. *Introduction to Finite Mathematics and Linear Programming*. Belmont, California: Wadsworth.

Simmons, D. M. 1972. *Linear Programming for Operations Research*. San Francisco, California: Holden-Day.

Spiveg, W. A., and R. M. Thrall 1970. *Linear Optimization*. New York: Holt, Rinehart and Winston.

Problems

1. Maximize $z = 10x_1 + 12x_2 + 8x_3$

subject to
$$4x_1 + 4x_2 \leqslant 10$$
$$5x_1 + 3x_2 + 4x_3 \leqslant 15$$
$$2x_1 + 2x_3 \leqslant 20$$
$$x_1 \geqslant 0 \quad x_2 \geqslant 0 \quad x_3 \geqslant 0$$

2. Maximize $z = 4x_1 + 5x_2 + 10x_3$

    subject to $20x_1 - 40x_2 + 50x_3 \leqslant 500$

    $40x_1 + 5x_2 - 90x_3 \leqslant 1000$

    $32x_1 + 25x_2 - 50x_3 \leqslant 400$

    $x_1 \geqslant 0 \qquad x_2 \geqslant 0 \qquad x_3 \geqslant 0$

3. Maximize $z = 2x_1 + 3x_2 + x_3 + 5x_4$

    subject to $2x_1 + x_3 + x_4 = 50$

    $4x_2 + x_3 + 2x_4 = 100$

    $x_1 + 2x_2 + 4x_4 \leqslant 80$

    $x_1 \geqslant 0 \qquad x_2 \geqslant 0 \qquad x_3 \geqslant 0 \qquad x_4 \geqslant 0$

4. Maximize $z = 3x_1 + 4x_2 + x_3 + 5x_4$

    subject to $x_1 + x_2 + 4x_3 \leqslant 20$

    $2x_2 + 5x_3 + x_4 \leqslant 40$

    $2x_1 + 6x_3 + 4x_4 \leqslant 50$

    $x_2 + x_3 + 3x_4 \leqslant 30$

    $x_1 \geqslant 0 \qquad x_2 \geqslant 0 \qquad x_3 \geqslant 0 \qquad x_4 \geqslant 0$

5. Minimize $z = -x_1 - x_2 - x_3$

    subject to $x_1 + x_2 + x_3 \leqslant 100$

    $x_2 + x_3 \leqslant 50$

    $x_1 \leqslant 50$

    $x_1 \geqslant 0 \qquad x_2 \geqslant 0 \qquad x_3 \geqslant 0$

6. Minimize $z = 4x_1 - 3x_2 + 5x_3 - 6x_4$

    subject to $-x_1 + 6x_2 \leqslant 100$

    $7x_2 - 4x_3 - 5x_4 \leqslant 200$

    $x_1 + 4x_2 + 5x_3 + 2x_4 \leqslant 400$

    $x_1, x_2, x_3, x_4 \geqslant 0$

7. Minimize $z = 3x_1 - 2x_2 - x_3$

    subject to $x_1 + 2x_2 + 3x_3 \geqslant 48$

    $x_1 + x_2 + x_3 \geqslant 50$

    $x_1 \geqslant 0 \qquad x_2 \geqslant 0 \qquad x_3 \geqslant 0$

8. Maximize $z = 0.25x_1 + x_2 + 2.50x_3$

   subject to $x_1 + 4x_2 + 2x_3 \leqslant 100$

   $$2x_1 \qquad + 8x_3 \leqslant 40$$

   $$x_1 \geqslant 0 \qquad x_2 \geqslant 0 \qquad x_3 \geqslant 0$$

9. Maximize $z = 2x_1 + x_2 + 5x_3 + 0.50x_4 + x_5$

   subject to $5x_1 + 3x_2 + 2x_3 + x_4 - x_5 \leqslant 32$

   $$x_1 + 2x_2 + 3x_3 + 4x_4 + 4x_5 \leqslant 10$$

   $$x_1 \geqslant 0 \quad x_2 \geqslant 0 \quad x_3 \geqslant 0 \quad x_4 \geqslant 0 \quad x_5 \geqslant 0$$

10. Solve Problem 2 of Chapter 2.

11. Solve Problem 5 of Chapter 2.

12. Solve Problem 6 of Chapter 2.

13. Solve Problem 8 of Chapter 2.

14. Show that a negative coefficient in the last row is a sufficient condition for optimality but not a necessary one. (Hint: Show that an optimal solution with a degenerate variable is identical to a tableau with a negative coefficient in the last row.)

15. Give conditions for which the tableau following a degenerate solution

    (a) will remain degenerate;
    (b) will not be degenerate.

16. What kind of solution exists if the coefficient in the last row is zero and no element in that column is positive?

17. What does it mean if the optimal solution contains an artificial variable that is basic but has a value of zero?

18. Solve the problem

    Minimize $z = -3x_1 + 8x_2 - 5x_3$

    subject to $-2x_3 \leqslant 5$

    $$2x_1 - 3x_2 + x_3 \leqslant 3$$

    $$2x_1 - 5x_2 + 6x_3 \leqslant 5$$

    $$x_1, x_2, x_3 \geqslant 0$$

    by converting the minimization problem to a maximization problem (Section 3.4).

19. Solve the problem

Minimize $z = 4x_1 - x_2 - x_3 + 2x_4 + x_5$

subject to $x_1 + x_2 + x_3 + x_4 + x_5 \leqslant 15$

$\qquad -x_1 + 2x_2 - x_3 + 2x_4 \qquad \leqslant 10$

$\qquad 2x_1 + x_2 - x_3 + x_4 - 2x_5 \leqslant 4$

$\qquad\qquad x_1, x_2, x_3, x_4, x_5 \geqslant 0$

using the optimality criteria for minimization problems (Section 3.4).

20. Using the simplex method, show that an unbounded solution exists for

Maximize $z = 5x + 4y$

subject to $0.10x + 0.05y \geqslant 40$

$\qquad 0.04x + 0.20y \geqslant 70$

$\qquad 0.06x + 0.10y \geqslant 60$

$\qquad\qquad x, y \geqslant 0$

Four

Linear Programming— Advanced Topics

4.1. Introduction

In Chapter 3 we examined the simplex method, which can be used to solve any linear programming problem. The simplex method requires that the coefficients in the constraints (a_{ij}), the amount of resources (b_i), and the cost coefficients (c_j) be known. In real-life applications, however, it may be necessary to vary any of these parameters for a variety of reasons: Fewer resources may be available than originally anticipated, the profit per unit may increase, the amount of resource i used for product j may change, a new constraint or a new product may be imposed on the problem. Rather than solve the problem again for every one of these events, we can use the information in the final tableau, as the starting tableau for the new problem. Doing this eliminates the need for artificial variables in order to find an initial solution. In this chapter the method for using the final tableau in order to accommodate the necessary changes in the problem is presented. Furthermore, any linear programming problem can be converted to a different problem. This new problem is termed the *dual* and its relationship to the original linear programming problem is both interesting and useful.

4.2. Adding New Constraints

Suppose that after performing some or all of the iterations of the simplex method, the decision maker realizes that a constraint is missing from the problem. As indicated in Chapter 3, this might well be the case if the solution to the problem is found to be unbounded. It is certainly possible to add the new constraint to the original problem and then solve the problem by restarting the simplex method at the beginning. However, a shortcut is available for adding the constraint to the final tableau; in this section we explain this shortcut after first illustrating the problem of adding a new constraint graphically.

Consider the Exclusive Furniture Company problem from Chapter 2:

Maximize $\qquad\qquad\qquad\qquad 5x+7y$

subject to $\qquad 2x+5y\leqslant 50\qquad$ materials constraint

$\qquad\qquad\quad\ 3x+2y\leqslant 42\qquad$ man-hour constraint

$\qquad\qquad\quad\ x\geqslant 0\quad y\geqslant 0\qquad$ nonnegativity restriction

The optimal solution is $x=10$, $y=6$ (10,6), which occurs at the intersection of the two constraint lines. Suppose that after solving the problem the Exclusive Furniture Company receives an order from a distributor for at least 14 pieces, at least 7 of which must be tables. Since the Exclusive Furniture Company has a policy of satisfying its customers whenever possible, it must add to the problem the two constraints

$\qquad\quad x+y\geqslant 14\qquad$ total number of pieces

and

$\qquad\qquad y\geqslant 7\qquad$ guaranteeing at least 7 tables

Figure 4.1. Three-Constraint Feasible Region

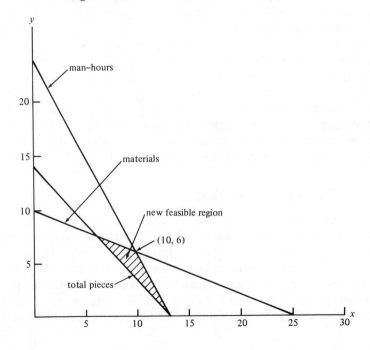

Figure 4.1 represents the original problem with the additional constraint on the total number of pieces. (The second new constraint is added later) Notice that the inclusion of the first new constraint reduces the size of the feasible region and that the new feasible region is contained within the old feasible region. Also notice that the point $(10,6)$ is still a feasible solution. Since $(10,6)$ is optimal for the large feasible region, $(10,6)$ yields a value of the objective function that is at least as large as the value of any point in the large feasible region. This includes all of the points in the new feasible region! Hence $(10,6)$ yields a value at least as good as any point in the new feasible region, and since it is a feasible solution it is therefore the optimal solution to the new problem. In fact, the following generalization can be made for any mathematical programming problem: Let R_1 be a feasible region and let x optimize a function $f(\cdot)$ on the region R_1. Then if R_2 is contained in R_1 and x is in R_2, x optimizes $f(\cdot)$ on the smaller region R_2. This is illustrated in Figure 4.2. Intuitively, we can sense the following: A new restriction cannot improve the objective function. Thus, if the objective function remains the same, it is optimal.

In Figure 4.3 the second constraint has been added to the problem. Again the size of the feasible region has been reduced. This time, however, $(10,6)$ is not contained in the feasible region. Hence a new optimal solution must be found, since $(10,6)$ cannot be optimal; in fact it is not even feasible. Thus, in this case the addition of a constraint has changed the solution to the problem. The new solution must have a value no larger than the 92 given by $(10,6)$, since the optimal value in the smaller region cannot be better than the optimal value in the larger region. Again, it follows that for any

Figure 4.2. Optimization on Smaller Feasible Region

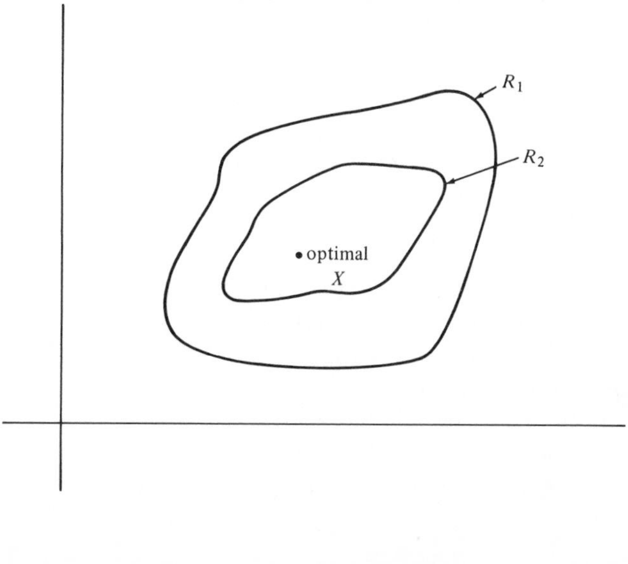

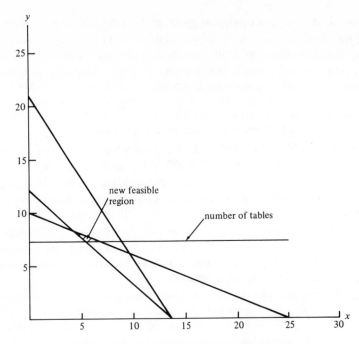

Figure 4.3. Four-Constraint Feasible Region

Figure 4.4. Reduced Feasible Region—New Optimal Solution

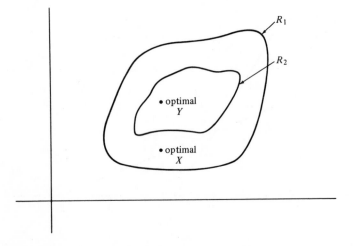

mathematical programming problem if R_1 is a feasible region and x optimizes $f(\cdot)$ on R_1, then if R_2 is contained in R_1 and y optimizes $f(\cdot)$ on R_2, $f(y)$ is no better than $f(x)$. This is illustrated in Figure 4.4.

In general, when adding a constraint to a mathematical programming problem there are three possible outcomes:

1. The constraint does not affect the optimal solution;
2. The constraint changes the optimal solution and
   (a) a new optimal solution can be found, or
   (b) the problem has no solution at all.

Examples of outcomes 1 and 2a are given above. An example of outcome 2b is adding the constraint $y \geqslant 12$ to the original problem. This is depicted in Figure 4.5.

From the graphical discussion it is apparent that the first step when adding a new constraint is to check whether the old optimal solution satisfies or violates the new constraint. If the constraint is satisfied, then the old optimal solution remains optimal. If the old optimal solution violates the new constraint, then the second step is to make the appropriate changes

Figure 4.5. No Feasible Region

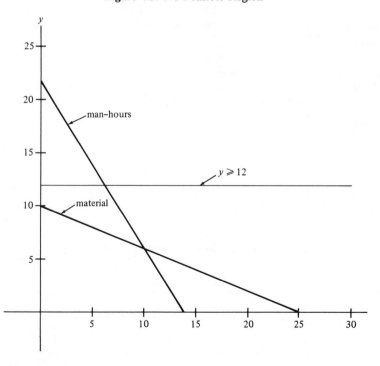

Table 4.1. Exclusive Furniture Company, Two constraints—Final tableau

c_b	Basic variables	Value	x	y	s_1	s_2
7	y	6	0	1	3/11	−2/11
5	x	10	1	0	−2/11	5/11
		92	0	0	1	1

in the final tableau and step 3 is to perform more iterations generating new tableaus and solutions. Consider now step 2.

Table 4.1 contains the final simplex tableau for the Exclusive Furniture Company problem given in Section 3.3. In order to add the constraint $x+y \geqslant 14$ to the problem, the constraint $x+y-s_3 = 14$ must be added to the final tableau where s_3 is a surplus variable. (For the moment, forget about the artificial variable.) The new tableau is given in Table 4.2. Notice that in addition to adding a row, we have added a column for s_3 and the values in the new column have been set to zero except in row 3. Also notice that this tableau is not in proper canonical form with respect to the basic variables y, x, and s_3. The tableau must be transformed so that the columns labeled y, x, and s_3 have all 0's except for a 1 in the row in which they are named as basic.

The only violations of these requirements occur in the new row, where -1 appears in column s_3 instead of $+1$ and 1's appear in columns x and y instead of 0's. Treat these problems one at a time, using the standard pivot operations of the simplex method. Correct the tableau by performing the operations on row 3 that would be necessary if each of the variables y, x, and s_3 had just entered the basis.

If s_3 had just entered the basis, then the pivot element would be column 5 (labeled s_3) and row 3. We divide row r by a_{rs} or in this case divide row 3 by -1. The new row 3 is given by

$$s_3 \quad -14 \quad -1 \quad -1 \quad 0 \quad 0 \quad 1$$

and none of the other rows have changed. Now suppose that y has just entered the basis in row 1. Then for $r=1$ and $s=2$ step 3 of the pivot

Table 4.2. Exclusive Furniture Company, Three constraints—First tableau

c_b	Basic variables	Value	x	y	s_1	s_2	s_3
7	y	6	0	1	3/11	−2/11	0
5	x	10	1	0	−2/11	5/11	0
0	s_3	14	1	1	0	0	−1
		92	0	0	1	1	0

operation must be performed on every element of row 3. That is,

$$\text{new } a_{3j} = \text{former } a_{3j} - \frac{a_{32}a_{1j}}{a_{12}}$$

or each new element in row 3 equals the corresponding old element minus a_{32}/a_{12} times the element in the same column of row 1. Since $a_{32}/a_{12} = -1/1 = -1$, the new row 3 = former row 3 $-(-1)$(row 1). In fact, in general step 3 of the pivot operation can be written as

$$\text{new row } i = \text{former row } i - (a_{is}/a_{rs})(\text{old row } r).$$

The pivot operation simply adds multiples of one row to another in order to produce basic columns of all zeros but one! Continuing with the example, we have new row 3 = old row 3 + row 1; hence the new row 3 is given by

$$s_3 \quad -8 \quad -1 \quad 0 \quad 3/11 \quad -2/11 \quad 1$$

Now treat x as a variable that just entered the basis in row 2. Hence $r=2$, $s=1$ and new row 3 = old row 3 $-(-1/1)$(row 2) or

$$\text{new row } 3 = \text{old row } 3 + \text{row } 2$$

yielding a new row 3 given by

$$s_3 \quad 2 \quad 0 \quad 0 \quad 1/11 \quad 3/11 \quad 1$$

Each of the basic variables is treated as if it had just entered the basis and the appropriate pivoting operations are performed on only the new row. Hence, the row that is added to the final tableau is given above and the new tableau is given in Table 4.3.

Notice that this last tableau is in canonical form and that the last row has all nonnegative values. Thus, this tableau is in the form of an optimal tableau and the current solution is optimal. That is, the optimal solution is $x=10$, $y=6$, $s_3=2$, $s_1=0=s_2$, and $z=92$. The addition of this new constraint has not changed the optimal solution to the problem which agrees with the graphical analysis.

Before considering the second constraint, let us summarize the necessary steps for adding a new constraint to the final tableau. Notice that the new row is numbered row $m+1$ and the new slack or surplus variable is numbered variable $n+m+1$.

Table 4.3. Exclusive Furniture Company, Three constraints—Final tableau

c_b	Basic variables	Value	x	y	s_1	s_2	s_3
7	y	6	0	1	3/11	−2/11	0
5	x	10	1	0	−2/11	5/11	0
0	s_3	2	0	0	1/11	3/11	1
		92	0	0	1	1	0

Table 4.4. Exclusive Furniture Company, Four constraints—First tableau

c_b	Basic variables	Value	x	y	s_1	s_2	s_3	s_4
7	y	6	0	1	3/11	−2/11	0	0
5	x	10	1	0	−2/11	5/11	0	0
0	s_3	2	0	0	1/11	3/11	1	0
0	s_4	−1	0	0	3/11	−2/11	0	1
		92	0	0	1	1	0	0

1. For the new constraint write the row as it would appear in the first tableau.
2. New $\text{row}_{m+1} = \text{old row}_{m+1}/a_{m+1,n+m+1}$.
3. For each row i, $i = 1, 2, \ldots, m$, let

$$\text{row}_{m+1} = \text{row}_{m+1} - \frac{a_{m+1,s}}{a_{is}} (\text{row } i)$$

where s is the column number of the basic variable in row i.
4. Add the final row_{m+1} to the tableau.
5. Add zeros to column $n+m+1$ in all places except row $m+1$.

Now consider the constraint $y \geqslant 7$. The new row to be added is

$$s_4 \quad 7 \quad 0 \quad 1 \quad 0 \quad 0 \quad 0 \quad -1$$

The first operation is to divide the row by -1 to obtain

$$s_4 \quad -7 \quad 0 \quad -1 \quad 0 \quad 0 \quad 0 \quad 1$$

The second operation is to subtract -1 times row 1 from the new row to obtain

$$s_4 \quad -1 \quad 0 \quad 0 \quad 3/11 \quad -2/11 \quad 0 \quad 1$$

The last two operations are to subtract 0 times row 2 and 0 times the last row; hence, the row added to the tableau is the row above and the tableau is as given in Table 4.4.

This time the tableau is not in proper canonical form, since $s_4 = -1$, which violates the restriction that all variables, including slack and surplus variables, be nonnegative. Somehow a move from an infeasible solution point to a feasible solution point is needed. The method for achieving this move is termed the *dual simplex method* and is presented next.

4.3. The Dual Simplex Method

The simplex method is a method that moves from feasible point to feasible point along a line, improving the objective function at each step. The dual simplex method is a method that moves from an infeasible point to an infeasible point along a line, reducing the objective function but attempting to find a feasible solution. The simplex method proceeds by choosing a variable to enter the basis and then determining which variable leaves the basis. The dual simplex method works in the opposite fashion. First a variable is chosen to leave the basis and then a variable is found to enter the basis. The simplex method stops when all of the cost coefficients are nonnegative or when it has been determined that the solution is unbounded. The dual simplex method stops when all of the variables are nonnegative or when it has been determined that having all nonnegative variables is impossible and no feasible solution exists. In short, the two methods are dual (opposite) to each other, whence the name dual simplex method.

Consider the last tableau of Section 4.2, given in Table 4.4. The problem with this tableau is that s_4 has a negative value. In order to alleviate this problem it might help if s_4 were to leave the basis. The immediate question is, which of the nonbasic variables should enter the basis? The graphical discussion in the preceding section indicates that the solution is "too good." Hence, the eligible variables for entering are those that *reduce* the value of the objective function, namely, the nonbasic variables with positive coefficients in the last row. The choice is between s_1 and s_2 and as with the simplex method the decision is made according to a ratio rule. The variable to enter the basis is given by

$$\frac{\bar{c}_s}{a_{rs}} = \underset{\substack{a_{rj}<0 \\ j=1,2,\ldots,n}}{\text{maximum}} \left\{ \frac{\bar{c}_j}{a_{rj}} \right\}$$

where $\bar{c}_s$ is the coefficient in column s of the last row. Notice that the ratios are taken only if a_{rj} is negative. In this example a_{43} is positive $(3/11)$ and a_{44} is negative $(-2/11)$, hence column 4 (s_2) becomes basic. Let $r=4$ and $s=4$, and perform the standard pivot operation of the simplex method. The new tableau is given in Table 4.5. Notice that all the variables are now nonnegative and that all of the coefficients in the last row are nonnegative. Hence the current solution is optimal for the four-constraint problem. The new optimal solution is $x=15/2$, $y=7$ and z has now been reduced from 92 to 86.5. The dual simplex method and its relationship to the simplex method are summarized in Figure 4.6.

Table 4.5. Exclusive Furniture Company Four constraints— Final tableau

c_b	Basic variables	Value	x	y	s_1	s_2	s_3	s_4
7	y	7	0	1	0	0	0	-1
5	x	15/2	1	0	1/2	0	0	5/2
0	s_3	1/2	0	0	1/2	0	1	3/2
0	s_2	11/2	0	0	$-3/2$	1	0	$-11/2$
		173/2	0	0	5/2	0	0	11/2

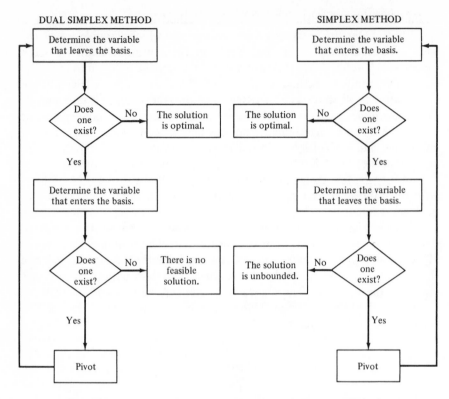

Figure 4.6. Comparison of Dual Simplex and Simplex Methods

The way the dual simplex method is used to add constraints should now be clear. However, the power of the dual simplex method is best illustrated by the diet problem of Section 2.4. The problem after formulation is

$$\text{Minimize} \qquad c = 0.02x + 0.03y$$

$$\text{subject to} \qquad 0.10x + 0.05y \geqslant 40$$

$$0.20x + 0.15y \geqslant 50$$

$$0.04x + 0.20\dot{y} \geqslant 70$$

$$0.10x + 0.10y \geqslant 10$$

$$0.06x + 0.10y \geqslant 60$$

$$x \geqslant 0 \qquad y \geqslant 0$$

According to the simplex method, a surplus and artificial variable must be added to each of the five constraints, yielding a problem of five constraints and twelve variables (two original variables, five surplus variables, and five artificial variables). Instead, we simply add surplus variables s_1 through s_5 and let these surplus variables yield the first solution. The initial tableau

Table 4.6. Diet Problem—Initial Dual Simplex Tableau

c_b	Basic variables	Value	x	y	s_1	s_2	s_3	s_4	s_5
0	s_1	−40	−0.1	−0.05	1	0	0	0	0
0	s_2	−50	−0.2	−0.15	0	1	0	0	0
0	s_3	−70	−0.04	−0.2	0	0	1	0	0
0	s_4	−10	−0.1	−0.1	0	0	0	1	0
0	s_5	−60	−0.06	−0.1	0	0	0	0	1
		0	0.02	0.03	0	0	0	0	0

then is as given in Table 4.6. Notice that minimize $c=0.02x+0.03y$ is converted to maximize $z=-0.02x-0.03y$ and that the initial solution is infeasible.

Using the dual simplex method we find that s_3 leaves the basis, since it is the most negative variable in the value column. The variable that enters the basis is determined by taking the maximum ratio of $\bar{c}_j/a_{rj}$ for $a_{rj}<0$. Since

$$\text{Maximum} \left\{ \frac{0.02}{-0.04}, \frac{0.03}{-0.20} \right\} = \text{Maximum} \{ -0.5, -0.15 \} = -0.15$$

the variable y enters the basis, s_3 leaves the basis, and a_{32} is the pivot element. The pivot operation is performed and the new tableau is given in Table 4.7. Notice that the value of the objective function is negative since the maximization algorithm is being used for a minimization problem. Also notice that not only has the negative variable s_3 left the basis but also s_2 and s_4 have become positive.

The dual simplex method has not ended because there still are negative values for two of the basic variables. Since s_5 is more negative than s_1, s_5 is chosen to leave the basis. The variable that enters is chosen according to the ratio test:

$$\text{Maximum} \left\{ \frac{0.014}{-0.04}, \frac{0.15}{-0.5} \right\} = \text{Maximum} \{ -0.35, -0.30 \} = -0.30$$

The maximum ratio occurs in the fifth column, which means that s_3 enters the basis and the pivot element is a_{55}. The tableau that results after pivoting

Table 4.7. Diet Problem—Second Dual Simplex Tableau

c_b	Basic variables	Value	x	y	s_1	s_2	s_3	s_4	s_5
0	s_1	−22.5	−0.09	0	1	0	−0.25	0	0
0	s_2	2.5	−0.17	0	0	1	−0.75	0	0
−0.03	y	350	0.20	1	0	0	−5	0	0
0	s_4	25	−0.08	0	0	0	−0.5	1	0
0	s_5	−25	−0.04	0	0	0	−0.5	0	1
		−10.5	0.014	0	0	0	0.15	0	0

Table 4.8. Diet Problem—Third Dual Simplex Tableau

c_b	Basic variables	Value	x	y	s_1	s_2	s_3	s_4	s_5
0	s_1	−10	−0.07	0	1	0	0	0	−0.5
0	s_2	40	−0.11	0	0	1	0	0	−1.5
−0.03	y	600	0.60	1	0	0	0	0	−10
0	s_4	50	−0.04	0	0	0	0	1	−1
0	s_3	50	0.08	0	0	0	1	0	−2
		−18	0.002	0	0	0	0	0	0.30

is given in Table 4.8. At this point, one basic variable remains negative, thus s_1 should leave the basis. The appropriate ratios are $0.002/-0.07$ for x and $0.30/-0.5$ for s_5. The larger of the two ratios occurs for x; hence, x enters the basis and the pivot element is a_{11}. The tableau after pivoting is given in Table 4.9. (The entries are rounded off to three decimal places.)

Since all basic variables are nonnegative, as are the coefficients in the last row, the problem is solved. Note that only three dual simplex iterations are required, whereas the simplex method would require at least five iterations (of a 6×12 tableau) just to force the artificial variables out of the basis. This example demonstrates that the dual simplex method is as efficient for solving linear programming problems of the form

$$\text{Minimize} \quad \sum_{i=1}^{n} c_j x_j$$

$$\text{subject to} \quad \sum_{i=1}^{n} a_{ij} x_j \geqslant b_i \qquad (4.1)$$

as the simplex method is for solving problems of the form

$$\text{Maximize} \quad \sum_{j=1}^{n} c_j x_j$$

$$\text{subject to} \quad \sum_{j=1}^{n} a_{ij} x_i \leqslant b_i \qquad (4.2)$$

Table 4.9. Diet Problem—Final Dual Simplex Tableau

c_b	Basic variables	Value	x	y	s_1	s_2	s_3	s_4	s_5
−0.02	x	142.857	1	0	−14.286	0	0	0	7.143
0	s_2	55.714	0	0	−0.571	1	0	0	−0.714
−0.03	y	514.286	0	1	−1.571	0	0	0	−14.572
0	s_4	55.714	0	0	0.571	0	0	1	−0.714
0	s_3	38.571	0	0	1.143	0	1	0	−2.571
		−18.286	0	0	0.003	0	0	0	0.286

In fact, it is no coincidence that the dual simplex method works well on Problem (4.1) because there is an intimate relationship between (4.1) and (4.2). This relationship is considered in the following section.

4.4. The Dual Problem

Consider, once more, the Exclusive Furniture Company problem:

$$\text{Maximize} \qquad 5x + 7y$$
$$\text{subject to} \qquad 2x + 5y \leqslant 50$$
$$3x + 2y \leqslant 42$$
$$x \geqslant \quad y \geqslant 0$$

Also, consider a problem that uses the same numbers but is expressed in a different manner:

$$\text{Minimize} \qquad 50u + 42v$$
$$\text{subject to} \qquad 2u + 3v \geqslant 5$$
$$5u + 2v \geqslant 7$$
$$u \geqslant 0 \quad v \geqslant 0$$

Notice that in the second problem the objective function and the right-hand side of the original problem have been interchanged; we are to minimize rather than maximize; the signs are greater than or equal to rather than less than or equal to; and the constraint coefficients that were rows are now columns. The original problem is termed the *primal*, while the new problem is termed the *dual*. Since the dual problem is of the form of Problem (4.1), the dual simplex method can be used to solve it. The results of the iterations are given in Table 4.10. The primal problem was solved in Section 3.3 and the iterations presented in Table 3.2. This table is adjoined to Table 4.10 in Table 4.11 so that the results of the simplex method on the primal and of the dual simplex method on the dual can be easily compared. Notice that the solution to each problem is 92. Note, too, that the numbers appearing in each tableau are the same (except that some signs are reversed). Also, the pivot elements at each step are the same except for sign reversal and, in fact, the ratios taken at each step are the same. Solving either the dual or the primal is equivalent to solving the other. All that needs to be done is to interpret the relationship between the two solutions.

In order to derive the dual problem, rows and columns of the primal have been transposed. In the dual problem, the coefficients of u are 2 in the first constraint and 5 in the second constraint. However, the 2 and 5 originally

Table 4.10. Exclusive Furniture Company—Dual Problem

c_b	Basic variables	Value	u	v	s_1	s_2
0	s_1	-5	-2	-3	1	0
0	s_2	-7	-5	-2	0	1
		0	50	42	0	0
0	s_1	-11/5	0	-11/5	1	-2/5
-50	u	7/5	1	2/5	0	-1/5
		-70	0	22	0	10
-42	v	1	0	1	5/11	2/11
-50	u	1	1	0	-2/11	3/11
		-92	0	0	10	6

come from the coefficients of the first constraint in the primal. Therefore, somehow the variable u is directly related to the first constraint of the primal. Similarly, v is directly related to the second constraint of the primal. Now notice that the optimal dual solution is $u=1$, $v=1$ and that in the optimal primal tableau the coefficients of slack variables s_1 and s_2 are both 1. This is because the variables u and v represent the values of the first and second resources, respectively. The primal problem presents these values in the final tableau while the dual problem uses these values directly as the variables. The values represent the amount that the objective function will increase if one unit of the resource is added and are called *shadow prices*. In this example, if one more man-hour is added, the objective function will increase from 92 to 93.

Table 4.11. Exclusive Furniture Company—Primal and Dual Problems

	Primal							Dual					
c_b	Basic variables	Value	x	y	s_1	s_2	c_b	Basic variables	Value	u	v	s_1	s_2
0	s_1	50	2	5	1	0	0	s_1	-5	-2	-3	1	0
0	s_2	42	3	2	0	1	0	s_2	-7	-5	-2	0	1
		0	-5	-7	0	0			0	50	42	0	0
7	y	10	2/5	1	1/5	0	0	s_1	-11/5	0	-11/5	1	-2/5
0	s_2	22	11/5	0	-2/5	1	-50	u	7/5	1	2/5	0	-1/5
		70	-11/5	0	7/5	0			-70	0	22	0	10
7	y	6	0	1	3/11	-2/11	-42	v	1	0	1	5/11	-2/11
5	x	10	1	0	-2/11	5/11	-50	u	1	1	0	-2/11	3/11
		92	0	0	1	1			-92	0	0	10	6

The rules for changing from a primal problem to a dual are very simple:

1. Change maximize to minimize or vice versa.
2. Transpose the rows to columns.
3. Change inequalities from $\geqslant$ to $\leqslant$ or vice versa.

For example, the original diet problem (the primal) is

$$\text{Minimize} \qquad c=0.02x+0.03y$$
$$\text{subject to} \qquad 0.10x+0.05y\geqslant 40$$
$$0.20x+0.15y\geqslant 50$$
$$0.04x+0.20y\geqslant 70$$
$$0.10x+0.10y\geqslant 10$$
$$0.06x+0.10y\geqslant 60$$
$$x\geqslant 0 \qquad y\geqslant 0$$

Hence, the dual problem is

$$\text{Maximize} \qquad 40y_1+50y_2+70y_3+10y_4+60y_5$$
$$\text{subject to} \quad 0.10y_1+0.20y_2+0.04y_3+0.10y_4+0.06y_5\leqslant 0.02$$
$$0.05y_1+0.15y_2+0.20y_3+0.10y_4+0.10y_5\leqslant 0.03$$
$$y_1\geqslant 0 \qquad y_2\geqslant 0 \qquad y_3\geqslant 0 \qquad y_4\geqslant 0 \quad y_5\geqslant 0$$

There is one possible advantage to solving the dual rather than the primal. If the number of constraints is larger than the number of variables, then the dual problem has more variables than constraints and is easier to solve using the simplex method. Of course, an alternative is to use the dual simplex method on the primal problem (as has been done for the diet problem).

4.5. Sensitivity Analysis*

In the previous sections of this chapter it has been shown that adding a new constraint poses few problems in light of the availability of the dual simplex method. In this section, consideration is given to the problems that occur if a new variable is added or if changes in the coefficient of the original problem need to be made. Of course, in any one of these circumstances the problem could be solved again from the beginning. However, this is not

---

*The material in this section is more difficult than that in the other sections and may be omitted without loss of continuity.

necessary, since the final tableau contains all of the information required to make the changes. Although the notation in this section is more difficult than that in previous sections, the concepts and computations are straightforward.

4.5.1. Adding a New Variable

It is very conceivable that after solving a linear programming problem the decision maker will realize that a variable not previously considered should be included in the problem. As an example, consider the Pentex gas problem of Section 3.5, which is given by:

Minimize $\quad\quad\quad\quad\quad 7x + 9y$

subject to $\quad\quad 0.3x + 0.5y \geqslant 2000 \quad\quad$ high test gasoline

$\quad\quad\quad\quad\quad 0.7x + 0.3y \geqslant 3000 \quad\quad$ regular gasoline

$\quad\quad\quad\quad\quad\quad\quad\quad x, y \geqslant 0$

Convert this problem to a maximization problem with $\leqslant$ inequalities and add slack variables.

Maximize $\quad\quad\quad\quad\quad -7x - 9y$

subject to $\quad\quad -0.3x - 0.5y + s_1 \quad\quad = -2000$

$\quad\quad\quad\quad\quad -0.7x - 0.3y \quad\quad + s_2 = -3000$

$\quad\quad\quad\quad\quad\quad\quad x, y, s_1, s_2 \geqslant 0$

The solution using the dual simplex method is given in Table 4.12. Suppose that a new gasoline can be used in the mixture and that this new gasoline contains 40% high test and 60% regular and costs 8.5¢ per gallon. Thus, the column given by

$$\begin{matrix} & w \\ & \begin{pmatrix} -0.4 \\ -0.6 \\ 8.5 \end{pmatrix} \end{matrix}$$

must be added to the original problem. However, since operations have already been performed on all three of the rows, we cannot add this column to the final tableau but instead must add the column that appears after performing the same operations as were performed to go from the first tableau to the final tableau. Referring back to Table 4.12 we see that these operations were, in order:

1. Divide row 2 by $-7/10$;
2. Subtract $-3/10$ times row 2 from row 1;

Table 4.12. Pentex Gas Problem—Dual Simplex Method

c_b	Basic variables	Value	x	y	s_1	s_2
0	s_1	-2000	$-\dfrac{3}{10}$	$-\dfrac{5}{10}$	1	0
0	s_2	-3000	$-\dfrac{7}{10}$	$-\dfrac{3}{10}$	0	1
		0	7	9	0	0
0	s_1	$-\dfrac{5000}{7}$	0	$-\dfrac{26}{70}$	1	$-\dfrac{3}{7}$
-7	x	$\dfrac{30{,}000}{7}$	1	$\dfrac{3}{7}$	0	$-\dfrac{10}{7}$
		$-30{,}000$	0	6	0	10
-9	y	$\dfrac{25{,}000}{13}$	0	1	$-\dfrac{35}{13}$	$\dfrac{15}{13}$
-7	x	$\dfrac{45{,}000}{13}$	1	0	$\dfrac{15}{13}$	$-\dfrac{25}{13}$
		$-\dfrac{540{,}000}{13}$	0	0	$\dfrac{210}{13}$	$\dfrac{40}{13}$

3. Subtract 7 times row 2 from the last row;
4. Divide row 1 by $-26/70$;
5. Subtract $3/7$ times row 1 from row 2;
6. Subtract 6 times row 1 from the last row.

Performing the same operations on the new column yields the transformation

Step	0	1	2	3	4	5	6
row	w	w	w	w	w	w	w
1	$-4/10$	$-4/10$	$-1/7$	$-1/7$	$5/13$	$5/13$	$5/13$
2	$-6/10$ $\rightarrow$	$6/7$ $\rightarrow$	$6/7$ $\rightarrow$	$6/7$ $\rightarrow$	$6/7$ $\rightarrow$	$9/13$ $\rightarrow$	$9/13$
last	$17/2$	$17/2$	$17/2$	$5/2$	$5/2$	$5/2$	$5/26$

(In the latter part of this section an alternative is presented whereby no tracking of previous iterations is necessary.) Thus, add to the final tableau the last column above to obtain Table 4.13. Note that this tableau is in perfect canonical form with respect to the basic variables and that in the last row all of the nonbasic coefficients are nonnegative. Even with the addition of a new type of gas the optimal purchasing schedule is the same as before. Thus, it is possible to add a variable and not change the solution. Is it possible for the solution to change: The answer is yes, and is best demonstrated by considering the column

$$
\begin{matrix}
v \\
\begin{pmatrix} -0.6 \\ -0.4 \\ 8 \end{pmatrix}
\end{matrix}
$$

Table 4.13. Pentex Gas Problem—Final Tableau with Gas w

c_b	Basic variables	Value	x	y	s_1	s_2	w
-9	y	$\dfrac{25{,}000}{13}$	0	1	$-\dfrac{35}{13}$	$\dfrac{15}{13}$	$\dfrac{5}{13}$
-7	x	$\dfrac{45{,}000}{13}$	1	0	$\dfrac{15}{13}$	$-\dfrac{25}{13}$	$\dfrac{9}{13}$
		$-\dfrac{540{,}000}{13}$	0	0	$\dfrac{210}{13}$	$\dfrac{40}{13}$	$\dfrac{5}{26}$

representing gasoline v, which has more high test and more regular than gasoline y and costs 1¢ less. Obviously this gasoline is better than gasoline y and should replace gas y in the final solution. In fact, after the same six operations have been performed, the new column is given by the final column below.

$$
\begin{array}{ccccccc}
\text{Step} & 0 & 1 & 2 & 3 & 4 & 5 & 6 \\
& v & v & v & v & v & v & v
\end{array}
$$

$$
\begin{pmatrix} -6/10 \\ -4/10 \\ 8 \end{pmatrix} \rightarrow
\begin{pmatrix} -6/10 \\ 4/7 \\ 8 \end{pmatrix} \rightarrow
\begin{pmatrix} -3/7 \\ 4/7 \\ 8 \end{pmatrix} \rightarrow
\begin{pmatrix} -3/7 \\ 4/7 \\ 4 \end{pmatrix} \rightarrow
\begin{pmatrix} 15/13 \\ 4/7 \\ 4 \end{pmatrix} \rightarrow
\begin{pmatrix} 15/13 \\ 1/13 \\ 4 \end{pmatrix} \rightarrow
\begin{pmatrix} 15/13 \\ 1/13 \\ -38/13 \end{pmatrix}
$$

When this column is added to the tableau it is obvious that more simplex iterations must be performed, since the solution improves (decreases) if v enters the basis. In fact, comparing ratios, we see that y is the variable to leave the basis. After pivoting, the tableau is given in Table 4.14 and the optimal solution is $x=3333$ and $v=1667$ with a cost of 36,667. In short, when a new variable is added, either the solution remains the same or the new variable should enter the basis and more simplex iterations should be performed.

The only question left is, must we keep track of all of the pivoting operations in order to add a new column? Fortunately, the answer is no. The information that is needed to add a new column to the final tableau can be found in the final tableau.

Table 4.14. Pentex Gas Problem—Final tableau with Gas v

c_b	Basic variables	Value	x	y	s_1	s_2	v
-8	v	1667	0	0.867	-2.33	1	1
-7	x	3333	1	-0.067	1.33	-2	0
		-36667	0	2.533	9.33	6	0

The six operations performed on the rows in order to solve the original Pentex gas problem were listed earlier. If these operations are performed on the surplus columns of the original tableau the surplus coefficients in row 1 in the final tableau are $-35/13$ and $15/13$, respectively. Since the original tableau had all zeros in column s_1 except in row 1, in order to get row 1 in the final tableau all of the six operations must "compound" to some operation that includes

$$\text{final row } 1 = \text{original row } 1(-35/13) + \text{something}$$

Similarly, since the only way to have s_2 in row 1 be equal to $15/13$ is to have in effect added $(15/13)(\text{row } 2)$ of the original tableau to something to get row 1 in the final tableau. In fact, it must be that the final row 1 is given by

$$\text{final row } 1 = \text{original row } 1(-35/13) + \text{original row } 2(15/13)$$

or else s_1 in row 1 is not $-35/13$ or s_2 in row 1 is not $15/13$. Let us check our computations:

A	original row 1	-2000	$-\dfrac{3}{10}$	$-\dfrac{5}{10}$	1	0
B	original row 2	-3000	$-\dfrac{7}{10}$	$-\dfrac{3}{10}$	0	1
C	row 1 $\left(\dfrac{-35}{13}\right) A$	$\dfrac{70000}{13}$	$\dfrac{105}{130}$	$\dfrac{175}{130}$	$\dfrac{-35}{13}$	0
D	row 2 $\left(\dfrac{15}{13}\right) B$	$\dfrac{-45000}{13}$	$\dfrac{-105}{130}$	$\dfrac{-45}{130}$	0	$\dfrac{15}{13}$
	$C+D=$ final row 1	$\dfrac{25000}{13}$	0	1	$\dfrac{-35}{13}$	$\dfrac{15}{13}$

This agrees with the final tableau.

Similar results exists for row 2, namely, that the final row 2 is given by

$$(\text{original row } 1)(15/13) + (\text{original row } 2)(-25/13)$$

Checking again, we find that

A	original row 1	-2000	$-\dfrac{3}{10}$	$-\dfrac{5}{10}$	1	0
B	original row 2	-3000	$-\dfrac{7}{10}$	$-\dfrac{3}{10}$	0	1
C	row 1 $\left(\dfrac{15}{13}\right)$	$-\dfrac{30,000}{13}$	$-\dfrac{9}{26}$	$-\dfrac{15}{26}$	$\dfrac{15}{13}$	0
D	row 2 $\left(\dfrac{-25}{13}\right)$	$\dfrac{75,000}{13}$	$\dfrac{35}{26}$	$\dfrac{15}{26}$	0	$\dfrac{-25}{13}$
	$C+D=$ final row 2	$\dfrac{45,000}{13}$	1	0	$\dfrac{15}{13}$	$\dfrac{-25}{13}$

which again agrees with row 2 in the final tableau. Hence, surplus or slack variable i in row j represents the number of times that the original row i is added to row j in order to get the last tableau. For any of the rows 1 through m then

$$\text{row } j \text{(final tableau)} = \sum_{i=1}^{m} \text{(final) slack}_{ji}^{*} \text{(initial) row } i$$

(The asterisk indicates that the element is from the final tableau). Finally, the last row is given by

$$\text{final last row} = \sum_{i=1}^{m} \text{slack}_{ji}^{*} \text{ row } i + \text{(original) last row}$$

In terms of the individual elements the relationships are

$$a_{ij}^{*} = \sum_{k=1}^{m} s_{jk}^{*} a_{kj}$$

for rows $1, 2, \ldots, m$, and

$$a_{\text{last}, j}^{*} = \sum_{k=1}^{m} s_{jk}^{*} a_{kj} + a_{\text{last}, j}$$

for the last row.

Thus, in order to add a new column in the last tableau only the original column and the coefficients of the slack variable in the last tableau need to be used. In the example, the first new column considered is

$$\begin{array}{c} w \\ \begin{pmatrix} -0.4 \\ -0.6 \\ 8.5 \end{pmatrix} \end{array}$$

and the surplus coefficients are given in the final tableau by

	s_1	s_2
row 1	$-35/13$	$15/13$
row 2	$15/13$	$-25/13$
last row	$210/13$	$40/13$

Hence

$$w(\text{row } 1) = (-35/13)(-0.4) + (15/13)(-0.6)$$
$$= 140/130 - 90/130 = 5/13$$
$$w(\text{row } 2) = (15/13)(-0.4) + (-25/13)(-0.6)$$
$$= -60/130 + 150/130 = 9/13$$

and

$$w(\text{last row}) = (210/13)(-0.4) + (40/13)(-0.6) + 8.5$$
$$= -840/130 - 240/130 + 17/2$$
$$= -216/26 + 221/26 = 5/26$$

which agrees with the column that is generated by performing the operations one at a time. Note that if this column is added to the last tableau, all of the elements in the last row are still greater than zero; hence the optimal solution does not change. Now consider the column generated by gas v.

$$\begin{array}{cc} v & v \\ \text{original} & \text{final} \\ \begin{pmatrix} -6/10 \\ -4/10 \\ 8 \end{pmatrix} & \rightarrow \quad \begin{pmatrix} 15/13 \\ 1/13 \\ -38/13 \end{pmatrix} \end{array}$$

Hence, adding this column would imply that v must enter the basis. It then is necessary to continue with the simplex method until the new optimal solution is found.

In summary, to add a new variable, list the old column $x_{n+1,1}, x_{n+1,2}, \dots,$ and the new column appended to the last tableau is given by

$$x_{n+1,j}(\text{new}) = \sum_{k=1}^{m} x_{n+1,k} a_{kj}^{*} \qquad \text{for } j = 1, 2, \dots, m$$

$$x_{n+1,\text{last}} = \sum_{k=1}^{m} x_{n+1,k} a_{kj}^{*} + x_{n+1,\text{last}}$$

At this point either an optimal solution is given by the current tableau or more simplex iterations have to be performed.

4.5.2. Sensitivity with Respect to the Resources

In all of the simplex method iterations the same operations are performed on the value column as on all other columns. Thus if the initial value column is changed from one to another, the numbers of the new value column in the final tableau are found by using the same method as in the previous section. For example, if after finding the optimal solution to the Exclusive Furniture Company problem the original right-hand side of the constraint equations is changed to

$$\begin{pmatrix} 55 \\ 44 \\ 0 \end{pmatrix}$$

instead of the original

$$\begin{pmatrix} 50 \\ 42 \\ 0 \end{pmatrix},$$

then in the last tableau the value column is given by

$$r_1 = (3/11)(55) + (-2/11)(44) = 15 - 8 = 7$$
$$r_2 = (-2/11)(55) + (5/11)(44) = -10 + 20 = 10$$
$$r_3 = (1)(55) + (1)(44) + 0 = 99$$

and the optimal solution is $y = 7$, $x = 10$, $z = 99$. Notice that it is possible for the new value column to have negative values, in which case the dual simplex method can be used. In this case, however, the value column has all nonnegative coefficients; hence the basis (x, y) is the same although the optimal values of x and y have changed, as has the value of the objective function. In short, a new right-hand side is handled the same way as adding a new column. The dual simplex method may be needed to solve the revised problem.

It is possible to determine how sensitive the profit is to a change in the amount of a particular resource. Suppose the original amount of the first resource, which is the raw material, is $50 + b$; then in the final tableau the values are

$$r_1 = (3/11)(50 + b) + (-2/11)(42) = 6 + (3/11)b$$
$$r_2 = (-2/11)(50 + b) + (5/11)(42) = 10 - (2/11)b$$
$$r_3 = (1)(50 + b) + 1(42) = 92 + b$$

Thus if b is positive, more of product y is produced and less of product x is produced until the amount of product x produced is 0 or $10 - (2/11)b = 0$, which means that if $b \geqslant 110/2$, the basis changes. Similarly, if b is negative, more of x and less of y are produced until $6 + (3/11)b = 0$ or $b = -66/3 = -22$. Hence, if the amount of resource 1 is between $50 - 22 = 28$ and $50 + 110/2 = 105$, the optimal solution is to produce both products in amounts $6 + (3/11)b$ and $10 - (2/11)b$, respectively, and the profit is $92 + b$. Similar results can be determined by varying the second resource, which is man-hours. Namely, let the amount of the second resource be $42 + c$; then

$$r_1 = 6 - (2/11)c \qquad r_2 = 10 + (5/11)c \qquad r_3 = 92 + c$$

Varying both simultaneously results in

$$r_1 = (3/11)(50 + b) - (2/11)(42 + c) = 6 + (3/11)b - (2/11)c$$
$$r_2 = (-2/11)(50 + b) + (5/11)(42 + c) = 10 - (2/11)b + (5/11)c$$
$$r_3 = (1)(50 + b) + 1(42 + c) = 92 + b + c$$

4.5.3. Changes in the Objective Function

In Section 4.5.1 we explained that any constraint row in the final tableau is represented as the sum of multiples of the original constraint rows; that is

$$\text{final row } i = \sum_{j=1}^{m} \text{slack}_{ij} \, (\text{original row } j)$$

where slack_{ij} comes from the final tableau. Notice that the last row, representing the objective function, does not enter into this computation. Hence, if the costs of the objective function change, the constraint rows are unaffected.

Now the last row in the final tableau is given by

$$\text{final last row} = \sum_{j=1}^{m} \text{slack}_{ij}\, \text{row}_j + \text{original last row}$$

Thus if a cost coefficient c_j changes from c_j to $c_j + b$, the final tableau has the identical change in the last row. For example, suppose that in the two-constraint Exclusive Furniture Company problem the profit per chair changes from 5 to 3 due to increased production costs. Then in the final tableau in column 1 of the last row we must add $+2$. This yields a final row given by

$$z \quad 92 \quad 2 \quad 0 \quad 1 \quad 1$$

The tableau is now not in canonical form since column 1 is not all 0's and one 1. In order to force a zero into this position, subtract 2 times row 2, the row in which x is basic, from the last row. Thus the last row becomes

$$z \quad 72 \quad 0 \quad 0 \quad 15/11 \quad 1/11$$

The tableau is now in canonical form and in fact it is in the form of an optimal tableau. Hence, the optimal solution is still $(10, 6)$; however, the profit of the problem with the revised cost is 72. In general, when changing costs there are two cases to consider for which the computations differ slightly if the new cost is for a basic or nonbasic variable. The procedure is as follows. If c_j changes to $c_j + b$, then

1. In the last row subtract b from column j;
2. If x_j is nonbasic, go to Step 4;
3. Add b times the row in which x_j is basic to the last row;
4. Examine the tableau for optimality. If optimal, stop. If not optimal, proceed with the simplex method, bringing in a new variable.

4.5.4. Changes in a Constraint Coefficient

Suppose that in the Exclusive Furniture Company problem the amount of man-hours necessary to produce a chair is reduced from 3 hours to 2 hours due to a new technology. Each row in the final tableau is the sum of a multiple of each initial row; therefore, changes must be performed in every row. However, since only column 1 changes, only column 1 needs to be updated. Thus, start by treating this as a problem with a new column. The column is given by

$$x$$
$$\begin{pmatrix} 2 \\ 2 \\ -5 \end{pmatrix}$$

Table 4.15. Exclusive Furniture Company—Change in Constraint Coefficient

c_b	Basic variables	Value	x	y	s_1	s_2
7	y	$\dfrac{8}{3}$	0	1	$\dfrac{1}{3}$	$-\dfrac{1}{3}$
5	x	$\dfrac{55}{3}$	1	0	$-\dfrac{1}{3}$	$\dfrac{5}{6}$
		$\dfrac{331}{3}$	0	0	$\dfrac{2}{3}$	$\dfrac{1}{6}$

and the final column is given by

$$\text{final } x_1 = (3/11)(2) + (-2/11)(2) = 6/11 - 4/11 = 2/11$$
$$x_2 = (-2/11)(2) + (5/11)(2) = -4/11 + 10/11 = 6/11$$
$$x_{\text{last}} = (1)(2) + (1)(2) + (-5) = 2 + 2 - 5 = -1$$

Hence, replace the final column 1 by

$$\begin{pmatrix} 2/11 \\ 6/11 \\ -1 \end{pmatrix}$$

Again the tableau is not in proper canonical form, since x is basic and the column should appear as all 0's and a 1 in row 2. Thus some of the pivot operations must be performed. These are the same as if row 2 had just been added to the tableau. Divide row 2 by $6/11$ and then add multiples of row 2 to row 1 and the last row. The final tableau is given by Table 4.15.

4.6. Summary

The analysis of linear programming is now complete. This chapter has provided the means to generalize linear programming problems so that it is possible to change coefficients or add constraints or variables. One of the useful methods for these changes is the *dual simplex method*, since this algorithm is capable of beginning with an infeasible solution and continuing until a feasible solution is found. The feasible solution is optimal and hence the dual simplex method is useful in its own right for problems that otherwise would require artificial variables.

The dual simplex method leads to the consideration of the *dual problem*. The *primal problem*, the original problem, has a related problem that is found by using three simple rules. Solving one problem is equivalent to solving the other because the primal and dual tableaus contain the same information. The dual variables are related to the primal constraints and in fact represent the *marginal cost* or *shadow prices* of the primal resources.

These shadow prices can be used to determine whether or not it is worthwhile to purchase additional resources.

Although it is possible to solve any linear programming problem using the simplex method or dual simplex method, there are certain types of problems that have a special structure which enables these problems to be solved by methods that are even more efficient than the simplex or dual simplex methods. The first of these problems, the transportation problem, is examined in the next chapter.

References and Selected Readings

Bradley, S. P., et al. 1977. *Applied Mathematical Programming*. New York: John Wiley and Sons, Inc.

Campbell, H. G. 1977. *Introduction to Matrices, Vectors and Linear Programming*. 2nd ed. Englewood Cliffs, New Jersey: Prentice-Hall, Inc.

Cooper, L., and D. Steinberg 1974. *Methods and Applications of Linear Programming*. Philadelphia: Saunders.

Gal, T. 1979. *Postoptimal Analyses, Parametric Programming and Related Topics*. New York: McGraw-Hill.

Gass, S. I. 1975. *Linear Programming*. 4th ed. New York: McGraw-Hill.

Simmons, D. M. 1972. *Linear Programming of Operations Research*. San Francisco: Holden-Day.

Swanson, L. W. 1979. *Linear Programming: Basic Theory and Applications*. New York: McGraw-Hill.

Problems

1. Given the problem

Maximize $z = 2x_1 + 2x_2$
subject to
$$x_1 \qquad\quad + x_3 \quad + x_4 \leqslant 1$$
$$x_2 \quad + x_3 \quad - x_4 \leqslant 1$$
$$x_1 \quad + x_2 \quad + 2x_3 \qquad\quad \leqslant 3$$
$$x_1 \geqslant 0 \quad x_2 \geqslant 0 \quad x_3 \geqslant 0 \quad x_4 \geqslant 0$$

Use the simplex method to find the solution and from the data in the optimal tableau establish whether the following additional constraints will change the optimal solution.
(a) $x_1 + x_2 \leqslant 1$.
(b) $x_1 - x_3 \geqslant 1$.

2. Solve the following linear programming problems by the dual simplex method.

(a) Minimize $z = 10x_1 + 5x_2 + 4x_3$
subject to $3x_1 + 2x_2 - 3x_3 \geqslant 3$
$$4x_1 \qquad\quad + 2x_3 \geqslant 10$$
$$x_1 \geqslant 0 \quad x_2 \geqslant 0 \quad x_3 \geqslant 0$$

(b)   Minimize $z = 3x_1 + 2x_2 + 4x_3$

   subject to $2x_1 - x_2 \quad \geqslant 5$

   $$2x_2 - x_3 \geqslant 10$$

   $x_1 \geqslant 0 \qquad x_2 \geqslant 0 \qquad x_3 \geqslant 0$

(c)   Minimize $z = 2x_1 + 4x_2 + 5x_3 + 3x_4$

   subject to $-x_1 - 2x_2 + 2x_3 \qquad \geqslant 40$

   $$3x_1 \qquad + 2x_3 + x_4 \leqslant 100$$

   $$x_1 - 2x_2 - x_3 + 4x_4 \geqslant 50$$

   $x_1 \geqslant 0 \qquad x_2 \geqslant 0 \qquad x_3 \geqslant 0 \qquad x_4 \geqslant 0$

3. Suppose a new variable, x_4, is introduced in Problem 2b. Solve the new problem for the following cost coefficient and activity vectors of x_4.

   (a) $c_4 = 1$ and $a_4 = \begin{pmatrix} -2 \\ 0 \end{pmatrix}$.   (b) $c_4 = 1$ and $a_4 = \begin{pmatrix} 0 \\ 2 \end{pmatrix}$.

   The activity vectors are given as they appear in the first dual simplex tableau.

4. Solve problem 2c, adding a new variable, x_5, with the following costs and activity vectors.

   (a) $c_5 = 2$ and $a_5 = \begin{pmatrix} -2 \\ 0 \\ -1 \end{pmatrix}$.

   (b) $c_5 = 2$ and $a_5 = \begin{pmatrix} 2 \\ 1 \\ -2 \end{pmatrix}$.

   Again, the a_5's are given as they appear in the tableau, not as in the original formulation.

5. Solve Problem 1 for each of the following modifications.

   (a) The new cost coefficient of x_2 is $c_2 = 4$.

   (b) The new resources vector is $\begin{pmatrix} 2 \\ 2 \\ 3 \end{pmatrix}$.

   (c) The new activity vector corresponding to x_2 is $\begin{pmatrix} 1 \\ 1 \\ 1 \end{pmatrix}$.

6. Show that the dual of the dual is the primal.

7. Show that an unrestricted variable in the primal corresponds to an equality constraint in the dual.

8. Solve

Minimize $\qquad 2x+3y$

subject to $\qquad 5x+2y \geqslant 6$

$\qquad\qquad\quad 2x-y \geqslant 3$

$\qquad\qquad\quad x, y \geqslant 0$

using the dual simplex method. Take the dual problem, apply the simplex method, and compare the two solution procedures.

9. Find examples for which both the primal and dual problems are infeasible.

10. Prove that changes in b will not affect a change in the basic variables, but may cause infeasibility.

Five

The Classical Transportation Problem

5.1. Background and Mathematical Formulation

The 1975 Nobel Prize for economics was awarded to Koopmann and Kantarovich for their independent development of the classical transportation problem. In 1939, Kantarovich (1960) formulated what was called the distribution problem, and in 1941 Koopmann (1951) developed the transportation problem, which happened to be equivalent to the distribution problem. However, Hitchcock (1941) was the first to publish the problem in a journal. Both Kantarovich and Koopmann suggested methods to solve the problem but their methods were inefficient. A breakthrough finally occurred in 1947 while Dantzig was working for the U.S. Air Force. He was instrumental in the formulation of an efficient algorithm for solving the transportation problem. This algorithm was later generalized by Dantzig in 1951 (see Dantzig, 1963), and it is commonly known as the linear programming problem.

Historically, the development of mathematical programming began with the transportation problem. The transportation problem is a specific case of linear programming, and any linear programming algorithm (for example, the simplex method) can be used to solve the transportation problem. When a general algorithm is applied to a special structural problem, it does not take advantage of the distinctive features in the problem and is therefore relatively ineffective. In short, it would be more advantageous to use an algorithm that is especially designed for solving a specific problem.

The classical transportation problem can best be described as one concerning the transportation of a single product from several sources to many destinations. Each source has a limited production capacity, and each destination has certain requirements regarding quantities of the product to be received. It should be recognized, however, that the production and marketing functions of the product are similar and production cost, quality, and selling prices are the same regardless of the origin or destination of the product. (This assumption is relaxed in Section 5.7.) The only factor that differs is the distribution cost, which primarily depends on the distance between source i and destination j. The task is then to minimize the overall transportation cost subject to the supply and demand restrictions for the product.

For the sake of discussion, it is now convenient to present an example concerning the Orlando Beer Company. The company produces beer at three breweries in Cleveland, Atlanta, and St. Louis, with plant capacities of 4, 6 and 8 million barrels per year, respectively. The beer is sold to five major beer distributors, located in New York, Philadelphia, Chicago, Milwaukee, and Boston, with an annual demand of 2, 4, 3, 4, and 5 million barrels, respectively. The data, including transportation costs are summarized in Table 5.1.

The cost of shipping 1 million barrels from Cleveland to New York is $c_{11} = \$6000$, while the cost of shipping the beer from Cleveland to Philadelphia is only $c_{12} = \$5000$. The matrix in Table 5.1 represents the costs c_{ij} in thousands of dollars of shipping from the three breweries to the five major distributors in the United States. The next step in the process is to formulate the descriptive problem mathematically by building a model, and then to solve it by a known technique.

Let s_i (in millions of barrels per year) represent the supply quantities at sources i:

$$i = 1 \qquad \text{Cleveland} \qquad s_1 = 4$$

$$i = 2 \qquad \text{Atlanta} \qquad s_2 = 6$$

$$i = 3 \qquad \text{St. Louis} \qquad s_3 = 8$$

Let d_j represent the demand (again, in millions of barrels per year) of the distributors at destinations j:

$$j = 1 \qquad \text{New York} \qquad d_1 = 2$$

$$j = 2 \qquad \text{Philadelphia} \qquad d_2 = 4$$

$$j = 3 \qquad \text{Chicago} \qquad d_2 = 3$$

$$j = 4 \qquad \text{Milwaukee} \qquad d_4 = 4$$

$$j = 5 \qquad \text{Boston} \qquad d_5 = 5$$

Let x_{ij} be the decision variable for brewery i and distributor j. Then the

Table 5.1. Orlando Beer Company Problem

From	To					Supply
	New York	Philadelphia	Chicago	Milwaukee	Boston	
Cleveland	$c_{11}=6$	$c_{12}=5$	$c_{13}=3$	$c_{14}=4$	$c_{15}=6$	4
Atlanta	$c_{21}=4$	$c_{22}=4$	$c_{23}=7$	$c_{24}=6$	$c_{25}=8$	6
St. Louis	$c_{31}=7$	$c_{32}=6$	$c_{33}=5$	$c_{34}=6$	$c_{35}=8$	8
Demand	2	4	3	4	5	

constraints in the problem are stated as in the accompanying table,

	Million barrels per year	Production capacity	Demand
$x_{11}+x_{12}+x_{13}+x_{14}+x_{15}$	$=4$	Cleveland	
$x_{21}+x_{22}+x_{23}+x_{24}+x_{25}$	$=6$	Atlanta	
$x_{31}+x_{32}+x_{33}+x_{34}+x_{35}$	$=8$	St. Louis	
$x_{11}+x_{21}+x_{31}$	$=2$		New York
$x_{12}+x_{22}+x_{32}$	$=4$		Philadelphia
$x_{13}+x_{23}+x_{33}$	$=3$		Chicago
$x_{14}+x_{24}+x_{34}$	$=4$		Milwaukee
$x_{15}+x_{25}+x_{35}$	$=5$		Boston

where $x_{ij} \geq 0$ $(i=1,2,3;\ j=1,2,3,4,5)$ indicates the amount of beer shipped from the ith brewery to the jth distributor. The objective function is to minimize the total transportation cost z, in thousands of dollars.

$$\text{Minimize } z = 6x_{11}+5x_{12}+3x_{13}+4x_{14}+6x_{15}+4x_{21}+4x_{22}$$
$$+7x_{23}+6x_{24}+8x_{25}+7x_{31}+6x_{32}+5x_{33}+6x_{34}+8x_{35}$$

In general, the transportation problem with m sources and n destinations can be expressed mathematically as
Find $x_{ij} \geq 0 (i=1,2,\ldots,m;\ j=1,2,\ldots,n)$ that

$$\text{Minimize} \qquad z = \sum_{i=1}^{m}\sum_{j=1}^{n} c_{ij}x_{ij}$$

$$\text{subject to} \qquad \sum_{j=1}^{n} x_{ij} = s_i \qquad i=1,2,\ldots,m \qquad (5.1)$$

$$\sum_{i=1}^{m} x_{ij} = d_j \qquad j=1,2,\ldots,n$$

It is important to note that in the numerical example, the problem is balanced in the sense that the total supply is equal to the total demand.

Furthermore, the assumption

$$\sum_{i=1}^{m} s_i = \sum_{j=1}^{n} d_j$$

remains valid throughout the chapter until Section 5.7, where unbalanced problems are solved.

Even though the problem is called the classical transportation problem, this technique can be used for solving problems that are not related to transportation—such as production planning, assignment, and plant location problems.

Like the simplex algorithm detailed in Chapter 3, the classical transportation algorithm consists of three steps. The first step is to determine a feasible starting solution (Section 5.2); the second is an optimality test (Section 5.3), which decides whether the procedure should stop or continue; and the third (Section 5.4) is a method for improving a nonoptimal solution.

5.2. Starting Solution Methods

The first step in solving a transportation problem is to come up with a starting solution. (Some algorithms require a feasible starting solution whereas others do not.) Four different methods that provide feasible starting solutions are presented in this chapter. (To be feasible, a solution must satisfy the constraints in (5.1).) The better the starting solution is, the fewer the number of improvements that are required to reach optimality. If we begin with a "bad" starting point, the computational time and effort are much longer than if we start with a good starting point. There is a trade-off between the amount of work invested in the starting solution and the time invested in improving the solution to optimality.

5.2.1. Northwest Corner

The first method for finding a starting solution is the simplest one because it does not consider transportation costs. For example, in Table 5.1, look at the "northwest" corner. The starting assignment from Cleveland to New York is at cell $(1,1)$ (the first number refers to the row and the second number to the column) and $x_{11} = 2$ is the lesser of the supply and demand of the two. The capacity in Cleveland is 4 but the demand in New York is 2. The difference between 4 and 2 is then shipped to Philadelphia, $x_{12} = 2$. Since Philadelphia is asking for $d_2 = 4$, and 2 units are arriving from Cleveland, the difference comes from Atlanta, $x_{22} = 2$. The balance of the table is completed by a similar process:

$$x_{11} = 2 \quad x_{12} = 2 \quad x_{22} = 2 \quad x_{23} = 3 \quad x_{24} = 1 \quad x_{34} = 3 \quad x_{35} = 5$$

and the remaining cells are empty. The starting solution can then be

Table 5.2. Northwest Corner Starting Solution

Brewery	Distributors				
	New York $d_1=2$	Philadelphia $d_2=4$	Chicago $d_3=3$	Milwaukee $d_4=4$	Boston $d_5=5$
Cleveland $s_1=4$	$c_{11}=6$ $x_{11}=2$	$c_{12}=5$ $x_{12}=2$	$c_{13}=3$	$c_{14}=4$	$c_{15}=6$
Atlanta $s_2=6$	$c_{21}=4$	$c_{22}=4$ $x_{22}=2$	$c_{23}=7$ $x_{23}=3$	$c_{24}=6$ $x_{24}=1$	$c_{25}=8$
St. Louis $s_3=8$	$c_{31}=7$	$c_{32}=6$	$c_{33}=5$	$c_{34}=6$ $x_{34}=3$	$c_{35}=8$ $x_{35}=5$

summarized as in Table 5.2. The cost of the starting solution is found by multiplying the unit costs c_{ij} times the volume x_{ij}:

$$z= \sum_{i=1}^{3} \sum_{j=1}^{5} c_{ij}x_{ij}=6\cdot2+5\cdot2+4\cdot2+7\cdot3+6\cdot1+6\cdot3+8\cdot5=\$115,000$$

The method just presented is simple and easy to apply. However, it does not give a good starting solution because the procedure does not consider the c_{ij}. Also, when assigning values x_{ij} you might want to know why we did not start from the northeast or southwest corners. As a matter of fact, any corner is an acceptable starting point and there is no way to tell in advance which one would provide the best starting point.

5.2.2. Minimal Column Value

The minimal column value method for finding a starting solution requires a few more computations than the northwest corner method, but it provides a better starting solution. This method requires the search of each column for the lowest cost, and the lowest cost determines the location from which the beer originates. The cost of shipping to New York from Cleveland, Atlanta, and St. Louis is $6, $4, and $7, respectively. Therefore, the lowest cost is from Atlanta, and it would be advantageous to ship as much as possible from this area. Since New York requires only 2, then $x_{21}=2$.

Philadelphia, the second destination, has costs of $5, $4, and $6 associated with it. Again, the lowest cost is from Atlanta, and therefore as much beer as possible should be shipped from Atlanta to Philadelphia, $x_{22}=4$. A similar analysis for Chicago yields $x_{13}=3$, shipped from Cleveland. The lowest possible way to ship to Milwaukee is from Cleveland, but since the decision was to ship 3 units from Cleveland to Chicago, only $x_{13}=1$ remains for Milwaukee. The other 3 for Milwaukee must arrive from St. Louis, $x_{34}=3$ (Atlanta is already committed), and the only possibility for Boston is to receive the beer from St. Louis, $x_{35}=5$. Therefore, the starting solution by the minimal column value method is

$$x_{21}=2 \quad x_{22}=4 \quad x_{13}=3 \quad x_{14}=1 \quad x_{34}=3 \quad x_{35}=5$$

This feasible solution costs

$$z= \sum_{i=1}^{3} \sum_{j=1}^{5} c_{ij}x_{ij}=4\cdot2+4\cdot4+3\cdot3+4\cdot1+6\cdot3+8\cdot5=\$95,000$$

It is clear that this solution is better than the northwest corner method solution. The solution, however, depends on the order in which the columns are chosen. A different order might yield a different starting solution.

An obvious variation of this method would be the search for the smallest cost value along each row rather than each column. Even though the procedure is basically the same, it is often more advantageous to use minimal column value if there are more columns than rows, and to use minimal row value if there are more rows than columns.

5.2.3. Minimal Matrix Value

The minimal matrix value method has characteristics that are similar to those of the minimal column value method. Instead of searching only for the lowest cost in each column, we search for the lowest cost in the entire matrix.

We set out to find the lowest cost in the matrix, which is $c_{13}=3$. The largest assignment from Cleveland to Chicago is $x_{13}=3$. The next smallest cost in the matrix is 4, which occurs in c_{21}, c_{22}, and c_{14}. If by chance a tie should occur, we can decide arbitrarily among all tied costs, and in this case c_{21} is chosen. The largest volume in c_{21} is $x_{21}=2$, followed by $x_{22}=4$, and $x_{14}=1$. The next smallest value in the matrix is $c_{12}=5$, but since Philadelphia has already received its requirement from Atlanta, we proceed to $c_{33}=5$ and find that Chicago has already been satisfied. The brewery that is still undetermined at this point is St. Louis, which requires shipping $x_{34}=3$ to Milwaukee and $x_{35}=5$ to Boston.

We can then proceed to list the starting solution as

$$x_{13}=3 \quad x_{14}=1 \quad x_{21}=2 \quad x_{22}=4 \quad x_{34}=3 \quad x_{35}=5$$

with the transportation cost of

$$z= \sum_{i=1}^{3} \sum_{j=1}^{5} c_{ij}x_{ij}=3\cdot3+4\cdot1+4\cdot2+4\cdot4+6\cdot3+8\cdot5=\$95,000$$

In this case, it is simply coincidental that the solution is the same as the one obtained by the minimal column value method.

5.2.4. Vogel's Approximation Method

The last method for finding a feasible starting solution is suggested by Vogel (Reinfeld and Vogel, 1958) and ˙ probably the best. (In fact in many cases the starting solution by the Vogel method is very close to or is the optimal one.) The basic idea is to assign penalties to a row whenever the lowest entry in that row is not utilized. The penalty is the difference between the lowest and the second-lowest transportation cost in the row. Thus, similar penalties can then be calculated for all rows and columns, where the column penalty is the difference between the lowest and second-lowest cost in a column. In order to avoid high penalties, the algorithm begins with the highest penalties, and assigns values to the lowest cost in that row or column.

The next numerical example demonstrates this procedure, and Table 5.3 includes the penalties for the rows and the columns. The penalty for Cleveland's shipping not to the lowest ($c_{13} = 3$) but to the second-lowest ($c_{14} = 4$) is $4 - 3 = 1$. Furthermore, the following penalties are obtained for

Atlanta	$c_{22} - c_{23} = 4 - 4 = 0$
St. Louis	$c_{32} - c_{33} = 6 - 5 = 1$
New York	$c_{11} - c_{21} = 6 - 4 = 2$
Philadelphia	$c_{12} - c_{22} = 5 - 4 = 1$
Chicago	$c_{33} - c_{13} = 5 - 3 = 2$
Milwaukee	$c_{24} - c_{14} = 6 - 4 = 2$
Boston	$c_{25} - c_{15} = 8 - 6 = 2$

The largest penalty is 2, and Boston is arbitrarily chosen from among the four cities with this penalty. In order to avoid this high penalty of 2, it is best to utilize the lowest channel leading to Boston, which is $c_{15} = 6$, arriving from Cleveland. The largest quantity that can be assigned there is $x_{15} = 4$, which is Cleveland's production level, and Boston still has a remaining demand of 1. The new penalties for the columns are now calculated accordingly. There is no need to calculate new penalties for rows, since there has been no change in the rows.

The new largest row or column penalty is 3 for New York. The lowest cost in the column is $c_{21} = 4$, and in order to avoid the penalty of 3 associated with New York, it is necessary to utilize the lowest cost in that

Table 5.3. Vogel's Approximation Method

Brewery	2 New York	4 Philadelphia	3 Chicago	4 Milwaukee	5 Boston	$(1)^a$ Penalty	$(4)^a$ Penalty	$(5)^a$ Penalty
					Distributor			
Cleveland 4	6	5	3	4	6 $x_{15}=4$	1		
Atlanta 6	4 $x_{21}=2$	4 $x_{22}=4$	7	6	8	0	2	2
St. Louis 8	7	6	5 $x_{33}=3$	6 $x_{34}=4$	8 $x_{35}=1$	1	1	0
Penalty $(2)^a$	2	1	2	2	2			
Penalty $(3)^a$	3	2	2	0	0			

^{a}The numbers in parentheses indicate the order of the penalty computations.

column. The amount that can be assigned is New York's demand thus $x_{21}=2$. It appears likely that the new row penalties should be calculated (what about the column penalties?). The highest penalty is now 2, and Chicago is arbitrarily chosen since c_{33} is the lowest cost in the column $x_{33}=3$, which clears Chicago and leaves St. Louis with capacity of 5. The new row penalties are calculated, and the next assignments to be considered are $x_{22}=4$, $x_{34}=4$, and $x_{35}=1$. Therefore, the starting feasible solution that is obtained is as follows:

$$x_{15}=4 \quad x_{21}=2 \quad x_{22}=4 \quad x_{33}=3 \quad x_{34}=4 \quad x_{35}=1$$

with total transportation cost of

$$z=6\cdot4+4\cdot2+4\cdot4+5\cdot3+6\cdot4+8\cdot1=\$95,000$$

Notice that the value of this solution is equal to the value of the last two starting solutions, again coincidentally.

5.3. Optimality Test

Once a starting solution is available, the next phase of the algorithm is a procedure that determines whether or not the solution at hand is optimal; this procedure is called an *optimality test*. Generally, all algorithms must have this feature, which either directs the search further or ends it. Two optimality tests are presented in this section, the stepping-stone and the MODI. The concept of either optimality test is to answer the question, does it pay to assign a numerical value to an empty cell? If the answer is yes, it is necessary to assign a numerical value and to provide a mechanism to maintain feasibility by not violating the constraints. However, if the answer is no, and empty cells should remain empty, this negative response indicates that an optimal solution has been reached and no further improvements are possible.

5.3.1. Stepping-Stone

Table 5.2 provides a feasible starting solution based on the northwest corner method. For this situation the solution is

$$x_{11}=2 \quad x_{12}=2 \quad x_{22}=2 \quad x_{23}=3 \quad x_{24}=1 \quad x_{34}=3 \quad x_{35}=5$$

$$x_{13}=x_{14}=x_{15}=x_{21}=x_{25}=x_{31}=x_{32}=x_{33}=0$$

We must now decide how to deal with the empty cells and determine which empty cells should stay empty and which ones should become positive cells. Specifically, we do not know whether x_{13} should remain at zero level or not. If we decide to increase the value of x_{13} from 0 to some positive level, for instance 1, then the same amount, 1, must be subtracted from x_{23}. The reason, of course, is that the third destination is asking for exactly $d_3=3$, and the assignments in the third column must add up to 3. By the same token, the supply at the first source is $s_1=4$, and the assignments in the first row must add to 4. If a positive amount is assigned to x_{13}, the same amount must be subtracted from x_{12} or x_{11}. Therefore, the subtraction of one unit from x_{23} has changed the balance in the second row, which must add to $s_2=6$, and the subtraction of a unit from x_{12} has changed the balance in the second column, which must add to $d_2=4$. The only way to resolve both situations is to add the same unit to x_{22}. The test for the empty cell x_{13} involves the following simultaneous changes:

Increase x_{13} by one unit;
Decrease x_{23} by one unit;
Decrease x_{12} by one unit;
Increase x_{22} by one unit.

These changes are a result of the feasibility constraints, and we must ascertain how much is gained, or lost, by making them. More specifically, the changes are

x_{13} an additional cost of $c_{13}=3$
x_{23} a saving of $c_{23}=7$
x_{12} a saving of $c_{12}=5$
x_{22} an additional cost of $c_{22}=4$

In short, the overall cost change is

$$c_{13}-c_{23}-c_{12}+c_{22}=3-7-5+4=-5$$

The suggested change indicates that x_{13} results in an additional cost of -5, or a saving of 5. Thus, increasing the value of x_{13} is a good strategy. If the cost resulting from the change is positive, then the empty cell should remain empty, since the objective is to minimize the overall transportation costs. The above process of finding the simultaneous changes around an empty cell is called finding a "loop" for this empty cell.

The empty cell $x_{13} = 0$ is not the only cell that should be evaluated. In fact, all empty cells should be evaluated, one at a time, and if no cell needs a change, then an optimal solution has been reached. The optimal solution occurs when every cell's evaluation is positive or zero and none is negative. Note further that each empty cell is evaluated by a loop as the process was described in previous paragraphs. This loop connects an empty cell (the one being evaluated) with some selected positive cells.

In keeping with the analogy of crossing a stream by stepping on stones, the positive cells can be thought of as stones. For a problem with m sources and n destinations where

$$\sum_{i=1}^{m} s_i = \sum_{j=1}^{n} d_j$$

there should always be $m+n-1$ positive cells and $m \cdot n - (m+n-1)$ empty cells. The number of positive cells $m+n-1$ results from the fact that there are m equations of the type $\sum_{j=1}^{n} x_{ij} = s_i$ and n equations of the type $\sum_{i=1}^{m} x_{ij} = d_j$, but since $\sum_{i=1}^{m} s_i = \sum_{j=1}^{n} d_j$ there is one redundant equation. Section 5.5 presents a problem that does not have the required number of positive cells.)

Whenever there are $m+n-1$ positive cells, it is always possible to find a loop around all empty cells. In fact, there is one and only one loop around each empty cell. Let us now return to the initial solution of the northwest corner presented in Table 5.2, and summarize the data in Table 5.4. The empty cell evaluation is always at the upper right-hand corner of the cell.

Table 5.4. Cell Evaluation

	$d_1=2$	$d_2=4$	$d_3=3$	$d_4=4$	$d_5=5$
$s_1=4$	6 / 2	5 / 2	3 −5	4 −3	6 −3
$s_2=6$	4 −1	4 / 2	7 / 3	6 / 1	8 0
$s_3=8$	7 2	6 2	5 −2	6 / 3	8 / 5

The results can also be summarized as follows:

Empty cell	Loop	Evaluation
(1,3)	(1,3), (1,2), (2,2), (3,2)	$c_{13} - c_{12} + c_{22} - c_{23} = 3 - 5 + 4 - 7 = -5$
(1,4)	(1,4), (1,2), (2,2), (2,4)	$c_{14} - c_{12} + c_{22} - c_{24} = 4 - 5 + 4 - 6 = -3$
(1,5)	(1,5), (1,2), (2,2), (2,4),	$c_{15} - c_{12} + c_{22} - c_{24} + c_{34} - c_{35}$
	(3,4), (3,5)	$= 6 - 5 + 4 - 6 + 6 - 8 = -3$
(2,1)	(2,1), (2,2), (1,2), (1,1)	$c_{21} - c_{22} + c_{12} - c_{11} = 4 - 4 + 5 - 6 = -1$
(2,5)	(2,5), (2,4), (3,4), (3,5)	$c_{25} - c_{24} + c_{34} - c_{35} = 8 - 6 + 6 - 8 = 0$
(3,1)	(3,1), (3,4), (2,4),	$c_{31} - c_{34} + c_{24} - c_{22} + c_{12} - c_{11}$
	(2,2), (1,2), (1,1)	$= 7 - 6 + 6 - 4 + 5 - 6 = 2$
(3,2)	(3,2), (3,4), (2,4), (2,2)	$c_{32} - c_{34} + c_{24} - c_{22} = 6 - 6 + 6 - 4 = 2$
(3,3)	(3,3), (3,4), (2,4), (2,3)	$c_{33} - c_{34} + c_{24} - c_{23} = 5 - 6 + 6 - 7 = -2$

The starting solution that results from the northwest corner method is so far away from the optimum that it is worth raising the values of almost all of the empty assignments to some positive level. In fact, at the moment only variables x_{25}, x_{31}, x_{32} should remain at level zero, since the evaluation of the variables is either zero or positive. However, an improvement of the other empty cells results in an improved total transportation cost.

As previously illustrated, it should always be possible to find a loop around the empty cells. All of the loops contain the empty cell being evaluated and either three or five positive cells. Other, more complicated loops may exist that contain an odd number of positive cells. Observe that each time a positive cell is reached in the process, a right or a left turn is made. Each time it is necessary to "step on a stone," the next stone is located by turning 90° right or 90° left. Furthermore, no row or column is entered more than once.

The stepping-stone method is a simple concept but may be long in the execution. Each cell is individually evaluated and many repetitions may be necessary until an optimal solution is achieved.

However, the stepping-stone method does provide an optimality test and can be useful. If all the empty cells are evaluated and one is found to be negative, an improvement is possible. If none is evaluated negatively, the solution at hand is an optimal solution.

5.3.2. Modified Distribution (MODI)

The stepping-stone method is a simple but a tedious way of determining whether or not a solution is optimal. It requires finding a loop for each empty cell and evaluating the cell. There are $m \cdot n - (m+n-1)$ empty cells, and for some of the empty cells, the loops are difficult to find. However, there is an alternative method for evaluating the empty cells without first determining the loops. This procedure is called the *modified distribution method* (MODI). The MODI technique provides information similar to that obtained by the stepping-stone method. This technique provides an algorithm that results in either an optimal solution or a way to improve the present solution.

The MODI method requires a new set of variables, u_i and v_j. Associated with each source (row) there is a variable u_i, and associated with each destination (column) there is a variable v_j. The m variables u_i and n variables v_j can be negative, positive, or zero.

The results of some duality theorems (see Chapter 4) applied to transportation problems, yield the following properties:

1. For all basic positive cells

$$c_{ij} - u_i - v_j = 0 \qquad i = 1, 2, \ldots, m \quad j = 1, 2, \ldots, n$$

   where c_{ij} are the given costs. There are $m+n$ unknowns (m u_i's and n v_j's) and there are $m+n-1$ positive cells (equations). Thus, the number of variables is one more than the number of equations and one variable can be chosen as a parameter and can be assigned an arbitrary value.

   Once the u_i and v_j are known, the following property provides an optimality test:

2. The current solution is optimal if $c_{ij} - u_i - v_j \geq 0$ for all empty cells. If for one empty cell, (i, j), $c_{ij} - u_i - v_j < 0$, the solution is not optimal. It can be improved by allocating some positive value to x_{ij}.

Now, let us apply the MODI technique in the following problem in order to clarify this method. (We shall return later to the example we have been working on, since the solution requires knowledge of degenerate solutions, which are discussed in Section 5.5).

A brick company has factories at A, B, and C, which supply bricks to four buildings D, E, F, G, which are currently under construction. The factory capacities are 200, 280, and 350 units, respectively, and the building requirements are 220, 240, 260, 110, respectively. The unit shipping costs are as follows:

	To			
From	D	E	F	G
A	4	5	7	1
B	6	7	3	2
C	2	1	4	3

The starting solution to this problem, using the minimal matrix value, is presented in Table 5.5. You are urged to verify the x_{ij} assignments according to an increasing order of c_{ij}.

$$x_{14} = 110 \quad x_{32} = 240 \quad x_{24} = 0 \quad x_{31} = 110 \quad x_{23} = 260 \quad x_{34} = 0$$
$$x_{11} = 90 \quad x_{33} = 0 \quad x_{12} = 0 \quad x_{21} = 20 \quad x_{22} = 0 \quad x_{13} = 0$$

Table 5.5. MODI Method

Factory capacity	Building requirements				
	$d_D=220$	$d_E=240$	$d_F=260$	$d_G=110$	
$s_A=200$	4 — 90	5 / 2	7 / 6	1 — 110	$u_1=0$
$s_B=280$	6 — 20	7 / 2	3 — 260	2 / -1	$u_2=2$
$s_C=350$	2 — 110	1 — 240	4 / 5	3 / 4	$u_3=-2$
	$v_1=4$	$v_2=3$	$v_3=1$	$v_4=1$	

Begin the search for u_i, v_j, by arbitrarily assigning $u_1=0$. The other values result from solving $c_{ij}-u_i-v_j=0$ for all positive cells in the following order:

Cell	Calculation	Result
(1,1)	$c_{11}-u_1-v_1=4-0-v_1=0$	$v_1=4$
(1,4)	$c_{14}-u_1-v_4=1-0-v_4=0$	$v_4=1$
(2,1)	$c_{21}-u_2-v_1=6-u_2-4=0$	$u_2=2$
(2,3)	$c_{23}-u_2-v_3=3-2-v_3=0$	$v_3=1$
(3,1)	$c_{31}-u_3-v_1=2-u_3-4=0$	$u_3=-2$
(3,2)	$c_{32}-u_3-v_2=1-(-2)-v_2=0$	$v_2=3$

There are ($m=3$ and $n=4$) $m+n-1=6$ equations, which enables us to solve for $u_2, u_3, u_4, v_1, v_2, v_3$; u_1 is arbitrarily chosen to equal zero.

The next step in the process is to evaluate all empty cells as to whether or not the optimality criteria are satisfied. More specifically, the test is to determine if $c_{ij}-u_j-v_j \geqslant 0$ for all empty cells (i,j).

Cell	Calculation	Optimality test
(1,2)	$c_{12}-u_1-v_2=5-0-3=2$	Satisfied
(1,3)	$c_{13}-u_1-v_3=7-0-1=6$	Satisfied
(2,2)	$c_{22}-u_2-v_2=7-2-3=2$	Satisfied
(2,4)	$c_{24}-u_2-v_4=2-2-1=-1$	Violated
(3,3)	$c_{33}-u_3-v_3=4-(-2)-1=5$	Satisfied
(3,4)	$c_{34}-u_3-v_4=3-(-2)-1=4$	Satisfied

After reviewing this example you might conclude that all $m \cdot n - (m+n-1) = 3 \cdot 4 - (3+4-1) = 12 - 6 = 6$ empty cells have been tested, and that five satisfied the optimality test and only one $(2,4)$ violated it. Therefore, the current solution is not optimal and could be improved by assigning a positive value to x_{24}. Note that it is convenient to put the empty cell's evaluation from the optimality test in the upper right-hand corner of each cell as in Table 5.5.

Once an empty cell with a negative cell evaluation has been located, it is important to remember that a loop around it does exist and can be determined as in the stepping-stone method. In this instance, the loop around $(2,4)$ is $(2,4)$, $(1,4)$, $(1,1)$, $(2,1)$. Notice that the cell evaluation for $(2,4)$ by the stepping-stone method is

$$c_{24} - c_{14} + c_{11} - c_{21} = 2 - 1 + 4 - 6 = -1$$

which is exactly equal to the value found by the MODI method. Moreover, all the values that are calculated for empty cells by the MODI method are equal to the cell evaluations obtained by the stepping-stone method. (You are urged to verify that this is so.) The difference, of course, is that the stepping-stone method requires determination of the loop before cell evaluation, whereas in the MODI method u_i and v_j are determined first. Based on the ability to evaluate all empty cells simultaneously, the MODI method is more efficient.

5.4. Improving a Nonoptimal Solution

The first two steps of any transportation algorithm involve finding a starting solution and an optimality test. The third and final component is a mechanism to improve a nonoptimal solution. It should be recognized that either an optimal solution or a negative cell evaluation is reached by the stepping-stone and MODI methods. If the solution is not optimal, there is a loop around an empty cell with a negative evaluation.

The next step in the process is to focus attention on that loop, since our aim is to improve the current solution. A negative evaluation for an empty cell indicates the amount of saving resulting from a one-unit increase in that empty cell. If a saving can be obtained by allocating one unit in an empty cell, then it also is advantageous to add more units to the cell. In fact, it is wise to increase the assignment by as much as possible whenever there is a negative cell evaluation. Along the loop, there are some previous assignments that will increase their values, while others are willing to decrease theirs. The latter cells, the ones that will decrease their value, determine the maximum possible amount that can be adjusted along the loop.

From among the cells whose assignment is going to decrease, choose the smallest volume for the desired quantity. Table 5.5 has one negative evaluation at $(2,4)$ with the loop around $(2,4)$ affecting the positive cells $(1,4)$, $(1,1)$, $(2,1)$. Because of the feasibility restriction, any amount added to

Table 5.6. Improved Solution

	$d_D=220$	$d_E=240$	$d_F=260$	$d_G=110$	
$s_A=200$	4 110	5 [2]	7 [5]	1 90	$u_1=0$
$s_B=280$	6 [1]	7 [3]	3 260	2 20	$u_2=1$
$s_C=350$	2 110	1 240	4 [4]	3 [4]	$u_3=-2$
	$v_1=4$	$v_2=3$	$v_3=2$	$v_4=1$	

$(2,4)$ must be subtracted from $(1,4)$ and $(2,1)$ and added to $(1,1)$. Therefore, the largest quantity that can be subtracted is 20. (If an increased amount is subtracted, cell $(2,1)$ becomes negative.) The new solution is illustrated in Table 5.6. The evaluation for cell $(2,4)$ is -1, indicating that the transportation cost decreases by 1 for each unit assigned to $(2,4)$. Since 20 units have been assigned to $(2,4)$, the total transportation cost is expected to decrease by 20 dollars. The total transportation cost for Table 5.5 is

$$4\cdot90+1\cdot110+6\cdot20+3\cdot260+2\cdot110+1\cdot240=1830$$

The total transportation cost for Table 5.6 is

$$4\cdot110+1\cdot90+3\cdot260+2\cdot20+2\cdot110+1\cdot240=1810$$

which is 20 dollars lower. The updated values for u_i, v_j that appear in Table 5.6 are calculated as follows. Arbitrarily assign $u_1=0$, which results in the following:

$$c_{11}-u_1-v_1=0 \Rightarrow v_1=4$$
$$c_{14}-u_1-v_4=0 \Rightarrow v_4=1$$
$$c_{24}-u_2-v_4=0 \Rightarrow u_2=1$$
$$c_{23}-u_2-v_3=0 \Rightarrow v_3=2$$
$$c_{31}-u_3-v_1=0 \Rightarrow u_3=-2$$
$$c_{32}-u_3-v_2=0 \Rightarrow v_2=3$$

With these u_i, v_j, we calculate the new cell evaluations, and the $c_{ij}-u_i-v_j$ are placed in the upper right-hand corner of the empty cells in Table 5.6. Thus, the present solution is an optimal one, since all the cell evaluations are positive.

Note that only one cycle is needed to reach an optimal solution for the problem. This is because the starting solution is a good one, since the calculations are performed by the minimal matrix value. However, we are not always so fortunate as to find such a good starting solution. If the starting solution is obtained by the northwest corner method, more cycles will be necessary.

To further exemplify this situation, consider a chewing gum company that manufactures gum at three locations 1, 2, 3, which have production capacities of 4, 5, 3 units, respectively. The gum is distributed to four locations 1, 2, 3, 4 with projected demands of 2, 4, 2, 4, respectively. The transportation cost of one unit is given as:

Destination

Source	$d_1=2$	$d_2=4$	$d_3=2$	$d_4=4$
$s_1=4$	6	5	3	4
$s_2=5$	4	4	7	6
$s_3=3$	7	6	5	6

In this case, the objective is to find the optimal distribution of gum from the three sources to the four destinations so that the overall transportation cost is minimized.

The starting solution as determined by the northwest corner method is given in Table 5.7. The order in which the positive values are assigned are

$$x_{11}=2 \quad x_{12}=2 \quad x_{22}=2 \quad x_{23}=2 \quad x_{24}=1 \quad x_{34}=3$$

With u_1 set equal to zero, the u_i, v_j will be calculated as:

$$u_1 = 0$$

$c_{11}-u_1-v_1=0$	$6-0-v_1=0$	$v_1=6$
$c_{12}-u_1-v_2=0$	$5-0-v_2=0$	$v_2=5$
$c_{22}-u_2-v_2=0$	$4-u_2-5=0$	$u_2=-1$
$c_{23}-u_2-v_3=0$	$7-(-1)-v_3=0$	$v_3=8$
$c_{24}-u_3-v_4=0$	$6-(-1)-v_4=0$	$v_4=7$
$c_{34}-u_3-v_4=0$	$6-u_3-7=0$	$u_3=-1$

The empty cells are then evaluated and the solution can be improved by increasing x_{14} through the loop x_{14}, x_{24}, x_{22}, x_{12}. (We cannot choose to increase x_{13} yet because although its cell evaluation is -5 and the loop around x_{13} could possibly decrease the cost more than a loop around x_{14}, such a choice results in a degenerate solution. This will be discussed in Section 5.5.) The order in which u_i, v_j are calculated in cycle 2 is

$$u_1=0 \quad v_1=6 \quad v_2=5 \quad v_4=4 \quad u_2=-1 \quad v_3=8 \quad u_3=2$$

Table 5.7. Chewing Gum Problem

Cycle 1 $z=68$

	$d_1=2$	$d_2=4$	$d_3=2$	$d_4=4$	
$s_1=4$	6 — 2	5 — 2_-	3 -5	4 -3 $+$	$u_1=0$
$s_2=5$	4 -1	4 — 2_+	7 — 2	6 — 1_-	$u_2=-1$
$s_3=3$	7 2	6 2	5 -2	6 — 3	$u_3=-1$
	$v_1=6$	$v_2=5$	$v_3=8$	$v_4=7$	

Cycle 2 $z=65$

	$d_1=2$	$d_2=4$	$d_3=2$	$d_4=4$	
$s_1=4$	6 — 2	5 — 1_-	3 -5 $+$	4 — 1	$u_1=0$
$s_2=5$	4 -1	4 — 3_+	7 — 2_-	6 3	$u_2=-1$
$s_3=3$	7 -1	6 -1	5 -5	6 — 3	$u_3=2$
	$v_1=6$	$v_2=5$	$v_3=8$	$v_4=4$	

Cycle 3 $z=60$

	$d_1=2$	$d_2=4$	$d_3=2$	$d_4=4$	
$s_1=4$	6 — 2_-	5 5	3 — 1_+	4 — 1	$u_1=0$
$s_2=5$	4 -6 $+$	4 — 4	7 — 1_-	6 -2	$u_2=4$
$s_3=3$	7 -1	6 4	5 0	6 — 3	$u_3=2$
	$v_1=6$	$v_2=0$	$v_3=3$	$v_4=4$	

125

Table 5.7 (*Continued*)

Cycle 4 ($z = 54$)

	$d_1 = 2$	$d_2 = 4$	$d_3 = 2$	$d_4 = 4$	
$s_1 = 4$	6 — 1_-	5 $\;-1$ — $+$	3 — 2	4 — 1	$u_1 = 0$
$s_2 = 5$	4 — 1_+	4 — 4_-	7 $\;6$	6 $\;4$	$u_2 = -2$
$s_3 = 3$	7 $\;-1$	6 $\;-2$	5 $\;0$	6 — 3	$u_3 = 2$
	$v_1 = 6$	$v_2 = 6$	$v_3 = 3$	$v_4 = 4$	

Cycle 5 ($z = 53$)

	$d_1 = 2$	$d_2 = 4$	$d_3 = 2$	$d_4 = 4$	
$s_1 = 4$	6 $\;1$	5 — 1_-	3 — 2	4 — 1_+	$u_1 = 0$
$s_2 = 5$	4 — 2	4 — 3	7 $\;5$	6 $\;3$	$u_2 = -1$
$s_3 = 5$	7 $\;0$	6 $\;-1$ — $+$	5 $\;0$	6 — 3_-	$u_3 = 2$
	$v_1 = 5$	$v_2 = 5$	$v_3 = 3$	$v_4 = 4$	

Cycle 6 ($z = 52$)

	$d_1 = 2$	$d_2 = 4$	$d_3 = 2$	$d_4 = 4$	
$s_1 = 4$	6 $\;2$	5 $\;1$	3 — 2	4 — 2	$u_1 = 0$
$s_2 = 5$	4 — 2	4 — 3	7 $\;4$	6 $\;2$	$u_2 = 0$
$s_3 = 3$	7 $\;1$	6 — 1	5 $\;0$	6 — 2	$u_3 = 2$
	$v_1 = 4$	$v_2 = 4$	$v_3 = 3$	$v_4 = 4$	

The solution is then improved through a loop on $x_{13}, x_{23}, x_{22}, x_{12}$ at cycle 2, which results in a better solution.

Note that cycle 5 can be omitted from the process. You can go directly from cycle 4 to cycle 6 by choosing the loop $x_{32}, x_{22}, x_{21}, x_{11}, x_{14}, x_{34}$. Unfortunately, *there is no way to predict which loop results in a fastest convergence to the optimal solution.*

As previously indicated, the number of cycles necessary to solve the problem depends greatly on the starting solution. The trade-off is between initial effort in securing a good starting solution and more work on improvements.

5.5. Special Cases

5.5.1. Alternative Optimal Solutions

From a practical perspective, it is important to know whether the optimal solution you obtain is unique or alternative optimal solutions exist. As with linear programming problems, there may be qualitative or nonquantitative considerations or other restrictions that are not included in the model. Consequently, it is important to provide the decision maker with more than one option in relation to a specific criterion.

In the transportation problem, it is easy to identify the alternative optimal solutions (similar to identifying alternative solutions in the linear programming process). Cell evaluation discloses how much more or less it costs the firm to place a positive value in the empty cell. A zero cell evaluation indicates that there is no change in the total transportation cost even though a value is placed in the cell.

In the example given in Table 5.7, the optimal solution is reached in the sixth cycle. It is important to point out that the cell evaluation of $(3,3)$ is zero. This means that the loop that passes through $(3,3)$, $(1,3)$, $(1,4)$, $(3,4)$ also has a value of zero. In this instance, the quantity that can be moved around the loop is 2 units, and the new alternative solution is then

	$d_1=2$	$d_2=4$	$d_3=2$	$d_4=4$
$s_1=4$	6	5	3	4 4
$s_2=5$	4 2	4 3	7	6
$s_3=3$	7	6 1	5 2	6

with $z=52$.

It is sufficient to look at the optimal solution: whenever there is a zero cell evaluation, there are alternative optimal solutions. If all cell evaluations are positive, then the optimal solution is unique.

5.5.2. Degenerate Solution

You may recall that little emphasis was given to degenerate solutions in our discussion of linear programming in Chapter 3, since the effects of degeneracy is minimal on linear programming. In the simplex tableau, the list of basic variables might include a degenerate variable with a value of zero. The situation is more complicated in transportation, however, since the names of all basic variables are not given. If one of the basic variables is zero, it is difficult to distinguish this variable from a nonbasic variable that is also zero.

There are two cases when a degenerate solution can cause difficulty. In the first instance, associated with the stepping-stone method, the attempt to find a loop around an empty cell is impeded. The second case involves the MODI method; in it, the effort to solve for the u_i, v_j is halted when the computation cannot be completed. In both cases the problem is a missing basic variable. It is essential to have $m+n-1$ basic variables, but if one of the variables is zero, the number of positive cells will only be $m+n-2$. An additional basic variable is zero, but determining its precise cell is difficult. A missing basic variable can prevent the completion of some loops. Also, the computation of u_i, v_j is based on $c_{ij} - u_i - v_j$ for *all* $m+n-1$ basic variables, so that if one is missing, an equation is missing.

A simple way to overcome these difficulties is to "tag" or "mark" the degenerate basic variable and to distinguish it from the other zero nonbasic variables. This is done by assigning a small positive amount, $\varepsilon > 0$, to an empty cell. Consider this amount to be small enough that it does not violate the capacity and demand restrictions. The degenerate variable should be positioned so that it enables us to find the loop or solve for u_i, v_j, neither of which could be done before.

We demonstrate this method by the example of the Orlando Beer Company given in Table 5.2. The initial solution is computed by the northwest corner method and Table 5.8 shows the iterations necessary to improve the initial solution.

Note that cycle 2 of Table 5.8 has $x_{12}, x_{21}, x_{23}, x_{24}, x_{34}, x_{35}$ as positive cells. The six basic variables are less than $m+n-1=3+5-1=7$, and one of the basic variables is zero. In determining u_i, v_j in cycle 2, it is possible to calculate $u_1 = 0, v_2 = 5$, but this is the end of the process. In order to complete the computations for u_i, v_j, assign $x_{22} = \varepsilon$. The total distribution in the second row is

$$x_{21} + x_{22} + x_{23} + x_{24} + x_{25} = 2 + \varepsilon + 3 + 1 + 0 = 6 + \varepsilon$$

and the second source has $s_2 = 6$. Since ε is small enough, it is possible to

128

Table 5.8. Orlando Beer Company Problem

Cycle 1 ($z=115$)

	$d_1=2$	$d_2=4$	$d_3=3$	$d_4=4$	$d_5=5$	
$s_1=4$	6; 2_-	5; 2_+	3 -5	4 -3	6 -3	$u_1=0$
$s_2=6$	4 -1; $+$	4; 2_-	7; 3	6; 1	8; 0	$u_2=-1$
$s_3=8$	7; 2	6; 2	5 -2	6; 3	8; 5	$u_3=-1$
	$v_1=6$	$v_2=5$	$v_3=8$	$v_4=7$	$v_5=9$	

Cycle 2 ($z=113$)

	$d_1=2$	$d_2=4$	$d_3=3$	$d_4=4$	$d_5=5$	
$s_1=4$	6; 1	5; 4_-	3 -5; $+$	4 -3	6 -3	$u_1=0$
$s_2=6$	4; 2	4; ε_+	7; 3_-	6; 1	8; 0	$u_2=-1$
$s_3=8$	7; 3	6; 2	5 -2	6; 3	8; 5	$u_3=-1$
	$v_1=5$	$v_2=5$	$v_3=8$	$v_4=7$	$v_5=9$	

Table 5.8 (*Continued*)

	$d_1=2$	$d_2=4$	$d_3=3$	$d_4=4$	$d_5=5$		
	6 1	5	3	4 −3	6 −3		
$s_1=4$		1_-	3	+		$u_1=0$	$z=98$
	4	4	7 5	6	8 0		
$s_2=6$	2	3_+		1_-		$u_2=-1$	
	7 3	6 2	5 3	6	8		
$s_3=8$				3	5	$u_3=-1$	Cycle 3
	$v_1=5$	$v_2=5$	$v_3=3$	$v_4=7$	$v_5=9$		

	$d_1=2$	$d_2=4$	$d_3=3$	$d_4=4$	$d_5=5$		
	6 2	5 1	3	4	6 0		
$s_1=4$			3	1		$u_1=0$	$z=95$
	4	4	7 4	6 2	8 2		
$s_2=6$	2	4				$u_2=0$	
	7 1	6	5 0	6	8		
$s_3=8$		ε		3	5	$u_3=2$	Cycle 4 Optimal solution
	$v_1=4$	$v_2=4$	$v_3=3$	$v_4=4$	$v_5=6$		

ignore the differences. The loop in cycle 2 calls for a move of 3 units around the loop $x_{13}, x_{23}, x_{22}, x_{12}$. Observe that by adding 3 units to ε in $(2,2)$, the ε is ignored. (The important role of ε is its usefulness as an indicator of a degenerate basic variable. Apart from this function, ε can be ignored.)

At cycle 4 of Table 5.8, the number of basic variables is 6, and when u_i, v_j are computed the result is

$$u_1 = 0 \qquad v_3 = 3 \qquad v_4 = 4 \qquad u_3 = 2 \qquad v_5 = 6$$

The ε now contributes an important function in estimating the other u_i, v_j. The location could be at any empty cell that has either known u_i and unknown v_j, or a known v_j and unknown u_i. In this instance, arbitrarily choose to put ε at $(3,2)$, which also happens to be an optimal solution. It is possible that a loop passing through ε would require subtracting a positive quantity from ε. The largest amount that can be subtracted from ε is ε itself, and that is the quantity that can be transferred around the loop. The practical interpretation of this procedure is that the location previously designated for ε is incorrect and it is necessary to move ε to a new location.

So far, the numerical examples of the transportation problems in this section have been structured so as to emphasize that one basic variable is equal to zero. However, some solutions could contain more than one degenerate variable. In such cases, the method for solving this problem is to assign as many ε's as necessary to generate $m + n - 1$ basic variables. When the degenerate solution has two basic variables that are zero, then two ε's are needed.

5.6. A Maximization Example

The transportation problems presented in the previous sections are examples for which the objective function is to be minimized. This section deals with transportation problems concerning c_{ij} as the profit of shipping one unit from source i to destination j. Obviously, the purpose here is to maximize total profit subject to the capacity and demand constraints. The algorithm is similar to the one for a minimization problem with minor modification to account for profit rather than cost. The differences can be summarized as follows.

1. In the stepping-stone and MODI methods, the solution can be improved if the cell evaluation is positive (not negative, as before) and the solution is optimal if the cell evaluations are negative or zero.
2. In the Vogel approximation method, the penalty is the difference between the highest profit and the second-highest profit (not the difference between the two lowest costs, as before).

Now consider the following problem concerning a dog food company. The L. Graham Company has a contract with four stores $1, 2, 3, 4$ to supply their dog food. The four stores' demands are 600, 800, 500, and 1300 pounds per week, respectively. The L. Graham Company has three plants A, B, C with production capacities of 1000, 400, and 1800 pounds per week, respectively. Based on the various wages, costs of raw material, and transportation costs, the company estimated the profit matrix of one pound of dog food as a function of its plants and stores.

		Store		
Plant	1	2	3	4
A	5	7	2	6
B	4	1	—	3
C	1	3	4	4

The profit of selling one pound produced at plant A and shipped to store 1 is 5¢; if shipped to store 2 it is 7¢; etc. Note that it is impossible to ship from plant B to store 3 because of a previous agreement between the company and the store. The Vogel approximation technique is applied to determine the starting solution and the computations are presented in Table 5.9. The penalties for the rows are

$$7-6=1 \qquad 4-3=1 \qquad 4-4=0$$

and the penalties for the columns are

$$5-4=1 \qquad 7-3=4 \qquad 4-2=2 \qquad 5-4=2$$

The data indicate that the largest penalty is 4, and it would be essential to avoid the penalty by assigning the quantity 800 to the largest profit in the second column x_{12}. The procedure continues, and the order of completing rows or columns is indicated by (1), (2), (3), (4).

The next step is to test whether or not the starting solution is optimal by using the MODI method. Since this is a maximization problem, the solution is optimal if the cell evaluations are negative or zero. Since there are positive values at $(1,1)(2,1)$, the solution can be improved. The loop in this situation is around $x_{11}, x_{14}, x_{34}, x_{31}$, and the quantity that is moved is 200 pounds. Therefore, the solution is improved from $z=\$134$ to $z=\$138$ and finally to $z=\$154$, which is the optimal solution (the optimal solution is a degenerate one).

5.7. Unbalanced Problems

Throughout this chapter only balanced problems are discussed, in that total supply is equal to total demand. In real-life situations, however, transportation problems may have either supply greater than demand or demand greater than supply. The former is best illustrated when production capabil-

Table 5.9. L. Graham Company Problem

Plant	d₁=600	d₂=800	d₃=500	d₄=1300	Penalty	Penalty	Penalty
	5	7	2	6			
$s_A=1000$		800		200	1	1	1 (4)
	4	1	—	3			
$s_B=400$				400	1	1	1
	1	3	4	4			
$s_C=1800$	600		500	700	0	0	3 (3)
Penalty	1	4 (1)	2 (2)	2			
Penalty	1			3			

Store

Plant	d₁=600	d₂=800	d₃=500	d₄=1300		
	5 | 3	7	2 | −4	6		
$s_A=1000$		800		200	$u_1=0$	$z=\$134$
	4 | 4	1 | −3	—	3		
$s_B=400$				400	$u_2=-3$	
	1	3 | −2	4	4		
$s_C=1800$	600		500	700	$u_3=-2$	Cycle 1
	$v_1=3$	$v_2=7$	$v_3=6$	$v_4=6$		

ity is greater than demand, and the latter situation occurs when a marketing requirement exceeds the production capacity. When supply exceeds demand, it is necessary to introduce a new artificial demand that absorbs the difference between total supply and demand. In the case of demand greater than supply, a dummy source is established to supply the difference between what is needed and what is available. The transportation cost from the dummy source to any destination is zero, since the artificial source does not really exist. Similarly, the transportation cost from any source to a dummy destination is zero for the same reason. By this simple modification process, an unbalanced transportation problem can be converted into a balanced one. The next step is to solve two problems, one with supply greater than demand, and the other with demand greater than supply.

Table 5.9 (*Continued*)

	Store					
Plant	$d_1=600$	$d_2=800$	$d_3=500$	$d_4=1300$		
$s_A=1000$	5 — 200	7 — 800	2 | −6	6 | −2	$u_1=0$	$z=\$138$
$s_B=400$	4 | 4	1 | −1	−	3 — 400	$u_2=-5$	
$s_C=1800$	1 — 400	3 | 0	4 — 500	4 — 900	$u_1=-4$	Cycle 2
	$v_1=5$	$v_2=7$	$v_3=8$	$v_4=8$		

	Store					
Plant	$d_1=600$	$d_2=800$	$d_3=500$	$d_4=1300$		
$s_A=1000$	5 — 200	7 — 800	2 | −4	6 — ε	$u_1=0$	$z=\$154$
$s_B=400$	4 — 400	1 | −5	−	3 | −2	$u_2=-1$	Cycle 3
$s_C=1800$	1 | −2	3 | −2	4 — 500	4 — 1300	$u_3=-2$	
	$v_1=5$	$v_2=7$	$v_3=6$	$v_4=6$		Optimal solution

5.7.1. Supply Greater Than Demand—Production Planning

The Baby-Bike Ltd. is planning its bicycle production for November, December, and January. Sales are 200 for November, increasing to 500 for December, and decreasing to 300 for January. The production level at Baby-Bike Ltd. is 250 bicycles per month with overtime capacity of an additional 150 bicycles. The marketing department has indicated a price of $20 a bicycle to the stores. The cost of production is $8 at regular time and $12 at overtime in each of the three months. However, each bicycle that is produced but not sold in the same month results in additional $5 inventory cost per month. (A $5 per month inventory cost is assigned for holding each bicycle for a month.) The decision variables for this case include the

Table 5.10. Production Planning

Sources	Destination			
	$j=1\ d_N=200$	$j=2\ d_D=500$	$j=3\ d_J=300$	$j=4$ dummy$=200$
$i=1\ SR_N=250$	12	7	2	0
$i=2\ SO_N=150$	8	3	-2	0
$i=3\ SR_D=250$	—	12	7	0
$i=4\ SO_D=150$	—	8	3	0
$i=5\ SR_J=250$	—	—	12	0
$i=6\ SO_J=150$	—	—	8	0

production level at regular and overtime hours for each of the three months, and the objective function is to maximize overall profit while satisfying demand restrictions.

Let SR_N, SR_D, SR_J represent the production capacity at regular time in November, December, and January, respectively, and SO_N, SO_D, SO_J the overtime production capacities during the three months. The transportation problem is then clearly illustrated in Table 5.10. The demand for November, December, and January is d_N, d_D, d_J, respectively, and it is now possible to build a profit matrix.

A bicycle that is produced in November during regular time costs $8 to produce, and if sold during November at $20 results in a profit of $20-8=$12$. However, if the same bicycle is produced in the overtime period, the cost is $12 and the profit is $20-\$12=\$8$, resulting in $c_{11}=\$12$, $c_{21}=\$8$. Moreover, if the same bicycles are stocked at the warehouse and sold during December, there is an additional holding cost of $5, which would further decrease the profit to $c_{12}=12-5=\$7$, $c_{22}=8-5=\$3$. There is also the possibility of having the bicycles stocked until January, which would reduce the profits further to $c_{13}=7-5=\$2$, $c_{23}=3-5=-\$2$ ($c_{23}=-\$2$ indicates that a bicycle produced in November at overtime and sold in January resulted in a loss of $2 to Baby-Bike Ltd). Bicycles produced and sold in December contribute a profit of $c_{32}=\$12$ and $c_{42}=\$8$, which is identical to the estimations for November. However, bicycles manufactured during December and stocked until January cost an additional $5 for inventory charges. Thus $c_{33}=12-5=\$7$ and $c_{43}=8-5=\$3$. Similarly, $c_{53}=12$ and $c_{63}=8$ are the only possibilities for January, since it is impossible to produce in January and sell in November or December. More specifically, the profit matrix is presented in Table 5.10. The total production capacity is

$$SR_N+SO_N+SR_D+SO_D+SR_J+SO_J=250+150+250+150+250+150$$
$$=1200 \text{ bicycles}$$

while demand is set at

$$d_N+d_D+d_J=200+500+300=1000 \text{ bicycles}$$

There is a need for a dummy destination of 200 that absorbs the difference between the supply of 1200 bikes and the demand for 1000 bikes; therefore, let dummy = 200. The profit of "selling" to the dummy destination is zero, regardless of the month or the factors of regular time or overtime, since this destination does not exist.

This problem can be solved by a regular transportation technique. The northwest corner method is used as a starting solution and the MODI method implemented for testing optimality. The iterations are presented in Table 5.11, and it is advantageous to review the procedure. After reviewing the example it is easy to see that the initial northwest corner solution and one iteration results in the optimal solution (it is a maximization problem and all cell evaluations are positive or zero).

Production Planning Schedule

	Produced	Sold
$x_{11} = 200$	November regular time	November
$x_{12} = 50$	November regular time	December
$x_{22} = 150$	November overtime	December
$x_{32} = 250$	December regular time	December
$x_{42} = 50$	December overtime	December
$x_{43} = 100$	December overtime	January
$x_{53} = 200$	January regular time	January
$x_{63} = 150$	January overtime	January

The total profit is

$$z = 12 \cdot 200 + 7 \cdot 50 + 3 \cdot 150 + 12 \cdot 250 + 8 \cdot 50 + 3 \cdot 100 + 12 \cdot 50 + 8 \cdot 150 = \$8700$$

This numerical example has alternative optimal solutions since there are several zero cell evaluations.

5.7.2. Demand Greater Than Supply

The Lucky Chain Company is located in Philadelphia and has four stores in the immediate vicinity. The company has a policy of planning televised commercials for each season, and three TV networks are available for the current summer period. Each store requires 12 minutes for commercials. The times available at the three networks are NCB, 10 minutes; ACB, 15 minutes; and CSB, 12 minutes. The cost of running a commercial varies from station to station, and from store to store (depending on production cost). The cost per minute in thousands of dollars is

Station	Store			
	A	B	C	D
NCB	3	3	2	5
ACB	2	1	3	2
CSB	5	3	4	4

Table 5.11. Production Planning Solution

	$j=1\ d_N=200$	$j=2\ d_D=500$	$j=3\ d_J=300$	$j=4$ dummy $=200$	
$i=1\ SR_N=250$	12 (200)	7 (50)	2 [Δ=0]	0 [Δ=10]	$u_1=0$
$i=2\ SO_N=150$	8 [Δ=0]	3 (150)	−2 [Δ=0]	0 [Δ=14]	$u_2=-4$
$i=3\ SR_D=250$	—	12 (250)	7 [Δ=0]	0 [Δ=5]	$u_3=5$
$i=4\ SO_D=150$	—	8 (50)	3 (100)	0 [Δ=9]	$u_4=1$
$i=5\ SR_J=250$	—	—	12 (200)	0 (50)	$u_5=10$
$i=6\ SO_D=150$	—	—	8 [Δ=−4]	0 (150)	$u_6=6$
	$v_1=12$	$v_2=7$	$v_3=2$	$v_4=-10$	

	$j=1\ d_N=200$	$j=2\ d_D=500$	$j=3\ d_J=300$	$j=4$ dummy $=200$	
$i=1\ SR_N=250$	12 (200)	7 (50)	2 [Δ=0]	0 [Δ=10]	$u_1=0$
$i=2\ SO_N=150$	8 [Δ=0]	3 (150)	−2 [Δ=0]	0 [Δ=14]	$u_2=-4$
$i=3\ SR_D=250$	—	12 (250)	7 [Δ=0]	0 [Δ=5]	$u_3=5$
$i=4\ SO_D=150$	—	8 (50)	3 (100)	0 [Δ=9]	$u_4=1$
$i=5\ SR_J=250$	—	—	12 (50)	0 (200)	$u_5=10$
$i=6\ SO_D=150$	—	—	8 (150)	0 [Δ=4]	$u_6=6$
	$v_1=12$	$v_2=7$	$v_3=2$	$v_4=-10$	

The problem is to find the minimal-cost TV commercial policy for the summer season for the Lucky Chain Company.

The initial tableau is illustrated in Table 5.12. The data further indicate that the total demand is $12+12+12+12=48$ minutes, while the total time available is $10+15+12=37$ minutes. Therefore, $48-37=11$ minutes must be available on a dummy television station with the costs of zero attached to each store on the station.

The initial solution as illustrated in Table 5.12 is also optimal, and further suggests:

TV Commercial Time

Store	ACB	CSB	NCB
A	12 minutes		
B	2 minutes	10 minutes	
C		2 minutes	10 minutes
D	1 minute instead of 12 required minutes		

This solution costs the company

$$2 \cdot 10 + 2 \cdot 12 + 1 \cdot 2 + 2 \cdot 1 + 3 \cdot 10 + 4 \cdot 2 = \$86,000$$

It may be further noted that an alternative solution does exist for this problem.

TV Commercial Time

Store	ACB	CSB	NCB
A	12 minutes		
B		12 minutes	
C			10 minutes
D	3 minutes		

Consequently, store C receives 10 from the required 12 minutes, and store D is allocated 3 of the 12 minutes that are necessary to fulfill their commercial requirement. However, there is another alternative solution presented in Table 5.12, with a different optimal solution. (For example, $x_{13}=10$, $x_{21}=3$, $x_{24}=12$, $x_{32}=12$, $x_{41}=9$, $x_{43}=2$.)

The manager of the Lucky Chain Company can now choose among several optimal solutions, one of which minimizes commercial cost for the summer. One criterion that is available to management is to equalize the commercials for each store. This was not a restriction in the original transportation problem.

Table 5.12. Lucky Chain Commercial Problem

	$A=12$	$B=12$	$C=12$	$D=12$	u_i	
NCB=10	3 1	3 2	2 10	5 3	0	
ACB=15	2 12	1 2_	3 1	2 +1	0	$z=86$
CSB=12	5 1	3 10+	4 −2	4 0	2	Starting solution
Dummy=11	0 0	0 1	0 + 0	0 −11	−2	Optimal solution
v_j	2	1	2	2		

	$A=12$	$B=12$	$C=12$	$D=12$	u_i	
NCB=10	3 1	3 2	2 10	5 3	0	$z=86$
ACB=15	2 12	1 0	3 1	2 3	0	Alternative optimal solution
CSB=12	5 1	3 12	4 0	4 ε	2	
Dummy=11	0 0	0 1	0 2	0 9	−2	
v_j	2	1	2	2		

Table 5.12 (*Continued*)

	A = 12	B = 12	C = 12	D = 12	u_i	
NCB = 10	3 | 1	3 | 2	2 10	5 | 3	0	$z = 86$
ACB = 15	2 3	1 | 0	3 | 1	2 12	0	Alternative optimal solution
CSB = 12	5 | 1	3 12	4 | 0	4 ε	2	
Dummy = 11	0 9	0 | 1	0 2	0 | 0	-2	
v_j	2	1	2	2		

5.8. Summary

The classical transportation problem concerns the need to ship goods from sources to destinations. Shipments may be real or conceptual as in the production planning example. The solution method has the same properties as that of the simplex method but is much simpler to execute. As usual, the first step is to find a *starting solution*, and there are several options for this first step, including *the northwest corner rule*, *the minimum column method*, *the minimum matrix method*, and *Vogel's approximation method*. An *optimality test* is applied to determine if the final solution has been achieved; if the solution is not an optimal one, either the *stepping-stone method* or the *MODI method* can be used to improve the current solution. The transportation algorithm clearly identifies the cases of degeneracy and alternative optimal solutions. Furthermore, if the problem is unbalanced, dummy sources or destinations can be added.

The next chapter discusses transportation problems that have an equal number of supply and destination points and each supply and each demand is exactly one. The problem is termed the assignment problem and the solution method is even simpler than that for the transportation problem.

References and Selected Readings

Cooper, L., and D. Steinberg 1974. *Methods and Applications of Linear Programming*. Philadelphia: Saunders.

Dantzig, G. B. 1963. *Linear Programming and Extensions*. Princeton, New Jersey: Princeton University Press.

Hillier, F., and G. Lieberman 1980. *Introduction to Operations Research*, 3d ed. San Francisco: Holden-Day.

Hitchcock, F. L. 1941. The distribution of a product from several sources to numerous localities. *J. Math. Phys.* 20:224–230.

Kantarovich. 1960. "Mathematical Methods in the Organization and Planning of Production," Publishing House of the Leningrad State University, 1939 translated in *Management Science*, Volume 6, pp. 366–422.

Koopmans, T. C. 1951. *Activity Analysis of Production and Allocation*. New York: John Wiley and Sons, Inc.

Reinfeld, N. V., and W. R. Vogel 1958. *Mathematical Programming*. Englewood Cliffs, New Jersey: Prentice-Hall, Inc.

Problems

1. The Sunco Distribution Company must meet the demand of five markets for solar heated swimming pools. Sunco has three warehouses at which limited amounts of the heating equipment are stored, and the firm can also buy unlimited amounts of the equipment from the manufacturer. The supplies, demands, and total costs (shipping and purchase cost) are given below. Find the optimal schedule.

Supply			
Warehouse	1	2	3
Number of units	600	500	400

Demand					
Market	1	2	3	4	5
Number of units	300	450	250	400	250

Total Costs Unit

	Markets				
	1	2	3	4	5
Warehouse 1	11.5	12.5	4.2	11.5	14
Warehouse 2	8	6	10	12.6	16
Warehouse 3	7	15	5.6	18	17
Manufacturer	24	30	35	25	26

2. Bush Wine Co. produces its product in plants in Atlanta, Baltimore, and Chicago that have weekly production capacities of 5000, 7000 and 4000 units, respectively. The plants must satisfy the requirements of distributors in the North, South, East, and West, and can also sell their weekly production

surplus in their local markets. The shipping costs, requirements, and selling price per unit are given in the table. Find the schedule that minimizes cost.

Plants	Markets				Selling price[a]
	North	South	East	West	
Atlanta	9.0	10.5	6.1	12.5	1.20
Baltimore	9.3	9.0	4.3	11.4	1.30
Chicago	9.5	4.3	4.5	9.8	1.16
Requirements	5000	4000	3800	2200	

[a] In dollars per unit.

3. A. Reagan Incorporated has warehouses in Los Angeles, San Francisco, San Diego, and Sacramento with 400, 500, 700, and 800 video cassettes, respectively. In the next month 200, 400, 200, 200, 100, and 300 cassettes must be shipped to six outlets. The shipping cost are given in the following table. Find the minimum cost shipping schedule.

Warehouses	Outlets					
	1	2	3	4	5	6
Los Angeles	8	6	10	12	9	11
San Francisco	3	7	6	9	8	7
San Diego	5	4	2	6	3	9
Sacramento	17	12	11	13	9	10

4. The Custom Costume company wants to allocate its production of four costumes among three plants. Sales are expected to be 3500, 2500, 2000, and 1500 units for Gorilla, Witch, Pumpkin, and Mork costumes, respectively. A plant can produce any combination of them and the capacity in units is 4000 for plant 1, 3500 for plant 2, and 4000 for plant 3. The company incurs a cost for unused capacity of $30 a unit at plant 1, $20 a unit at plant 2, and $15 a unit at plant 3. Given the following cost per unit of production, determine the minimum cost schedule.

Plant	Costumes			
	Gorilla	Witch	Pumpkin	Mork
1	30	15	20	18
2	10	12	22	9
3	25	22	18	15

5. Algonquin Airline Company is planning its operations for the coming year. The managers want to decide how to distribute the company fleet of 17 airplanes to serve the four territories in the country: northwest, southwest, northeast, and southeast. The fleet is composed of seven model CD 12 airplanes, four model CD 9, and six model CD 10. The manager feels that at least three planes should be assigned to each territory. Each of the planes can perform equally in any given area. In addition, a plane that is left unassigned

can be rented for private use. The profits in millions of dollars per year for each model and territory, as well as the estimate revenue for rental services are given in the table below. Find the optimal schedule.

	Northwest	Southwest	Northeast	Southeast	Rental
CD 12	3	7	7	5	4
CD 9	5	6	3	3	5
CD 2	2	4	4	2	3

6. Adscam Inc. has five secret distribution centers for its secret product. The centers are supplied by any of the company's secret plants. Each distribution center must receive 20 shipments a week. The distance in miles between plants and centers is given in the table. There is a fixed cost per shipment of $50 plus 50¢ per mile. The capacity of the plants is 40, 30, 20, and 10 shipments a week.

			Centers		
Plant	1	2	3	4	5
1	100	1000	350	200	1500
2	1000	1200	420	350	1400
3	800	1400	330	450	1100
4	600	1000	900	900	1100

Find the minimal cost shipment schedule.

7. Use the transportation simplex method to solve the following transportation problems. Determine the value of the objective function for both the primal and the dual problems.

(a) Maximize z

	$d_1 = 15$	$d_2 = 15$	$d_3 = 10$	$d_4 = 5$
$s_1 = 10$	3	1	5	2
$s_2 = 20$	4	3	4	1
$s_3 = 15$	5	2	6	3

(b) Minimize z

	$d_1 = 30$	$d_2 = 25$	$d_3 = 35$	$d_4 = 40$	$d_5 = 15$
$s_1 = 50$	5	6	4	10	5
$s_2 = 40$	2	8	3	12	4
$s_3 = 35$	1	10	2	11	6
$s_4 = 20$	3	10	4	10	7

(c) Minimize z

	$d_1 = 5$	$d_2 = 9$	$d_3 = 8$	$d_4 = 5$	$d_5 = 19$	$d_6 = 4$
$s_1 = 19$	10	9	8	10	11	8
$s_2 = 37$	11	12	7	8	10	7
$s_3 = 34$	8	10	5	12	10	6

(d) Minimize z

	$d_1=17$	$d_2=18$	$d_3=30$	$d_4=25$
$s_1=19$	5	3	6	2
$s_2=37$	4	7	9	1
$s_3=34$	3	4	7	5

(e) Minimize z

	$d_1=100$	$d_2=75$	$d_3=30$
$s_1=50$	2	2	3
$s_2=50$	3	4	2
$s_3=40$	1	1	1
$s_4=35$	3	1	1
$s_5=30$	2	2	4

(f) Minimize z

	$d_1=50$	$d_2=20$	$d_3=10$	$d_4=35$	$d_5=15$	$d_6=50$
$s_1=30$	5	10	15	8	9	7
$s_2=40$	14	13	10	9	20	21
$s_3=10$	15	11	13	25	8	12
$s_4=100$	9	13	12	8	6	13

8. Write the dual of the following transportation problem.

Minimize $z=3x_{11}+4x_{12}+6x_{13}+2x_{21}+4x_{22}+5x_{23}$

subject to x_{11} $+x_{12}$ $+x_{13}$ $=400$

x_{21} $+x_{22}$ $+x_{23}=300$

$x_{11}+x_{21}$ $=250$

$x_{12}+x_{22}$ $=300$

$x_{13}+x_{23}=150$

$x_{ij}\geqslant0$

9. Prove that adding a constant to any cost row does not change the optimal solution.

10. Why does Vogel's method not lead to an optimal solution every time? (Give the reasons.)

11. What is wrong with the statement, "If we cannot utilize the lowest cost we shall use the second-lowest cost?"

12. Exercise for Section 5.3.2:

		Distributors				
		New York	Philadelphia	Chicago	Milwaukee	Boston
Breweries		$d_1=2$	$d_2=4$	$d_3=3$	$d_4=5$	$d_5=5$
Cleveland	$s_1=4$	6	5	3	4	6
Atlanta	$s_2=6$	4	4	7	6	8
St.Louis	$s_3=8$	7	6	5	6	8

(a) Using the MODI method, evaluate the empty cells for the brewery problem discussed in Section 5.1.

(b) Is the current assignment optimal? Why or why not?

13. Using Table 5.7, begin at cycle 1 and increase x_{13} by 2 units through the loop $x_{13}, x_{23}, x_{22}, x_{21}$. Verify that this loop produces a degenerate solution using the ε technique discussed in this section.

(a) Complete the search for an optimal solution.

(b) Verify that the value of z in the optimal solution is the same as the value of z in the text's optimal solution.

14. Find an alternative optimal solution to Table 5.11.

Six

The Assignment Problem

6.1. Introduction

The assignment problem is a distinctive form of mathematical or linear programming that is similar to the transportation problem. The assignment problem is characterized by its simplicity (a one-to-one matching of two sets) and the efficient algorithm used to obtain a solution. Since the assignment problem is a special case of linear programming (Section 6.4), it is possible to apply either a simplex technique or a transportation method in order to solve it. While these two methods are viable techniques, they are not the most effective ones, since they do not utilize the unique structure of the assignment problem. A more efficient method of solving the assignment problem is presented in this chapter.

6.2. Description and Mathematical Statement of the Problem

The assignment problem can be described as one in which items are allocated to particular categories. Consider four persons—Adam, Ben, Carol, and Deb—who can perform any of four jobs, 1, 2, 3, and 4. The performance of the jobs varies from person to person, due to differences in education and skills. Table 6.1 indicates the cost associated with each person's performance on each of the four jobs. It is apparent that the cost associated with completing job 1 varies from $10 for Ben to $40 for Adam. The performance criterion is given in terms of cost, but it can be represented instead by time, efficiency, quality, or profit.

Other examples of assignment problems include

1. The allocation of a sales force to several districts;
2. A foreman scheduling jobs on an assembly line;
3. The assignment of teachers to classes;

Table 6.1. Assignment Costs

Person	Job			
	1	2	3	4
Adam	40	60	60	70
Ben	10	60	70	30
Carol	20	50	40	20
Deb	30	20	10	40

4. The bidding of n companies on m projects;
5. The distribution of raw materials to production areas.

An assignment problem can be defined as the one-to-one matching of two sets. For example, a computerized dating service has the task of matching a set of males with a set of females. There is a performance measure c_{ij} of matching male i with female j that can be either maximized or minimized.

A mathematical statement of the problem is similar to the one presented for the transportation problem. The decision variable x_{ij} is 0 if i is not assigned to j, and 1 if i is assigned to j. The objective function can then be formulated as

$$\text{Minimize} \qquad z = \sum_{i=1}^{n} \sum_{j=1}^{n} c_{ij} x_{ij}$$

$$\text{subject to} \qquad \sum_{i=1}^{n} x_{ij} = 1 \qquad j = 1, 2, \ldots, n$$

$$\sum_{j=1}^{n} x_{ij} = 1 \qquad i = 1, , \ldots, n$$

$$x_{ij} = 0 \text{ or } 1$$

The objective function is defined here as a minimization of the cumulative cost of the matches. The first constraint indicates that each male i must be matched to exactly one of the females, and the second constraint specifies that each female j must be matched to exactly one of the males.

Returning to the example presented in Table 6.1, we see that it is possible to enumerate all of the feasible permutations of assigning n persons to n jobs. After the possibilities have been listed and evaluated, a simple comparison yields the optimal solution. The first person can be assigned to one of n different tasks while the second person has $n-1$ jobs available. Obviously, the third person can choose from $n-2$ jobs; and the process continues until the last person is assigned the last job. The total number of combinations is

$$n(n-1) \cdot (n-2) \cdot \cdots \cdot 2 \cdot 1 = n!$$

As anticipated, $n!$ increases quickly and may be illustrated by

$$10! = 3.6 \times 10^6 \quad \text{and} \quad 15! = 1.3 \times 10^{12}$$

It should be recognized that there are $4! = 4 \cdot 3 \cdot 2 \cdot 1 = 24$ permutations for the problem in Table 6.1. These are listed in Table 6.2. The solution A1 B2 C3 D4 indicates that Adam is assigned to job 1, Ben to job 2, Carol to job 3, and Deb to job 4; the solution costs $40 + 60 + 40 + 40 = \$180$. The optimal solution, which is determined by the lowest cost of all solutions in Table 6.2, is

Person	Job	Cost
Adam	2	$60
Ben	1	10
Carol	4	20
Deb	3	10

The total cost of this assignment is $60 + 10 + 20 + 10 = \$100$. A complete enumeration performed manually is a feasible technique if the problem concerns the assignment of a relatively small number of items, say 4. If the number of items n is 8, the number of possible solutions is $8! = 6720$ and it

Table 6.2. Enumeration of all Solutions

Solution				Cost
A1	B2	C3	D4	$40+60+40+40=180$
A1	B2	C4	D3	$40+60+20+10=130$
A1	B3	C2	D4	$40+70+50+40=200$
A1	B3	C4	D2	$40+70+20+20=150$
A1	B4	C2	D3	$40+30+50+10=130$
A1	B4	C3	D2	$40+30+40+20=130$
A2	B1	C3	D4	$60+10+40+40=150$
A2	B1	C4	D3	$60+10+20+10=100 \leftarrow$ Optimal
A2	B3	C1	D4	$60+70+20+40=210$
A2	B3	C4	D1	$60+70+20+30=180$
A2	B4	C1	D3	$60+30+20+10=120$
A2	B4	C3	D1	$60+30+30+30=150$
A3	B1	C2	D4	$60+10+50+40=160$
A3	B1	C4	D2	$60+10+20+20=110$
A3	B2	C1	D4	$60+60+20+40=180$
A3	B2	C4	D1	$60+60+20+30=170$
A3	B4	C1	D2	$60+30+20+20=130$
A3	B4	C2	D1	$60+30+50+30=170$
A4	B1	C2	D3	$70+10+50+10=140$
A4	B1	C3	D2	$70+10+40+20=140$
A4	B2	C1	D3	$70+60+20+10=160$
A4	B2	C3	D1	$70+60+40+30=200$
A4	B3	C1	D2	$70+70+20+20=180$
A4	B3	C2	D1	$70+70+50+30=220$

is more advantageous to use the computer than to perform the task manually. However, even the computer is limited, and for large n, say $n \geqslant 20$, explicit enumeration is impractical, costly, and perhaps impossible. The Hungarian method, which solves large problems with relatively little effort, can be applied for large n and is presented in the next section.

6.3. Solution Using the Hungarian Method

It is easy to reach the optimal solution using the Hungarian method since the algorithm is simple and requires a minimal number of computations. It is named after the Hungarian mathematician D. Honig, who developed the original theorem required for the algorithm. The procedure is based on the following two properties of the assignment problem.

Property 1. Any constant may be added or subtracted from each element in a row or a column of the cost matrix without affecting the decision problem. Thus a reduction of the values in a row or a column by a constant changes the value of the optimal solution but does not alter the optimal decision variables x_{ij}. The assignment problem depends on the relative differences between row and column elements and not on their absolute values. If the costs are c_{ij} and k_i is subtracted from row i, then the cost of assigning person i in row j is now $c_{ij} - k_i$, regardless of where that person is assigned.

The mathematical proof of this property is presented in the following example. The fact that the cost coefficient is changed does not affect the feasibility of the problem; however, the value of the objective function is altered by it. The original objective function is

$$\sum_{i=1}^{n} \sum_{j=1}^{n} c_{ij} x_{ij}$$

Let us now subtract k_i from row i, which results in a new cost for row i of $(c_{ij} - k_i)$. The objective function is now

$$c_{11}x_{11} + c_{12}x_{12} + \cdots + c_{1n}x_{1n} + c_{21}x_{21} + c_{22}x_{22} + \cdots + c_{2n}x_{2n}$$
$$+ (c_{i1} - k_i)x_{i1} + (c_{i2} - k_i)x_{i2} + \qquad \cdots \qquad + (c_{in} - k_i)x_{in}$$
$$+ \cdots + c_{n1}x_{n1} + c_{n2}x_{n2} + \qquad \cdots \qquad + c_{nn}x_{nn}$$

$$= \sum_{i=1}^{n} \sum_{j=1}^{n} c_{ij} x_{ij} - k_i \sum_{j=1}^{n} x_{ij}$$

$$= \sum_{i=1}^{n} \sum_{j=1}^{n} c_{ij} x_{ij} - k_i \cdot 1$$

Since $\sum\limits_{j=1}^{n} x_{ij} = 1$ and k_i is a constant, the objective function has been altered by the constant k_i. This affects the value of the optimal solution but not the actual assignment.

Property 1 can also be applied to the transportation problem presented in Chapter 5. (You should now attempt to prove Property 1 for the transportation problem.)

The Hungarian method is based on the fact that a constant may be added to or subtracted from the rows or columns of the cost matrix.

Property 2. If the cost matrix contains only positive or zero values, the optimal solution has a total cost of zero or some positive number. More specifically, since the cost matrix contains only nonnegative costs, the solution is the summation of the nonnegative costs. If a solution of zero is feasible, then this zero solution is optimal.

The procedure for the Hungarian method. The Hungarian method assumes that the cost matrix contains only nonnegative elements. If this is not the case, a large positive constant is added to the rows in order to produce a nonnegative matrix. The formal steps of the algorithm are now presented.

Step 1a
Select the smallest element in each row and subtract it from each row element. Repeat this for all rows.

Step 1b
Select the smallest element in each column and subtract it from each column element. Repeat this procedure for the other columns.

Notice that at the end of Step 1a there is at least one zero element in each row, and at the end of Step 1b there is at least one zero element in each column.

In order to perform Step 1a in Table 6.1, subtract 40 from the first row, 10 from the second row, 20 from the third row, and 10 from the fourth row. This is illustrated as follows.

original table				subtraction		table at end of Step 1a		
40	60	60	70	-40	0	20	20	30
10	60	70	30	-10	0	50	60	20
20	50	40	20	-20	0	30	20	0
30	20	10	40	-10	20	10	0	30

In Step 1b it is not necessary to subtract 0 from the first, third, or fourth columns, but only 10 from the second column:

						table at end of Step 1b		
	0	20	20	30	0	10	20	30
	0	50	60	20	0	40	60	20
	0	30	20	0	0	20	20	0
	20	10	0	20	20	0	0	30
subtraction	−0	−10	−0	−0				

Step 2

Next cover the zeros with the smallest possible number of horizontal and vertical lines. If $n=4$ lines are necessary, an optimal solution has been reached. If less than four lines are used, proceed to Step 3.

Step 2 is verified by the following procedure. Determine the maximum number of possible zero assignments. This number is always equal to the number of minimal lines required to cover the zeros.

For example, the possible zero assignments are enclosed in a box in the following matrix:

$$\boxed{0} \quad 10 \quad 20 \quad 30$$
$$0 \quad 40 \quad 60 \quad 20$$
$$0 \quad 20 \quad 20 \quad \boxed{0}$$
$$20 \quad \boxed{0} \quad 0 \quad 30$$

Three zeros are assigned for $n=4$. Thus all zeros can then be covered by three lines, as follows:

$$0 \quad 10 \quad 20 \quad 30$$
$$0 \quad 40 \quad 60 \quad 20$$
$$0 \quad 20 \quad 20 \quad 0$$
$$20 \quad 0 \quad 0 \quad 30$$

Three lines (one horizontal and two vertical) are required to cover the zeros. For larger assignment problems, Step 2 is not easy. Furthermore, if at any iteration the number of lines is not equal to the number of possible zero assignments, this indicates either an arithmetic error, an error in locating the zero assignments, or an error in placing the covering lines. There are other possibilities for covering the zeros in this example with three lines; for example,

$$0 \quad 10 \quad 20 \quad 30$$
$$0 \quad 40 \quad 60 \quad 20$$
$$0 \quad 20 \quad 20 \quad 0$$
$$20 \quad 0 \quad 0 \quad 30$$

Step 3

For this step, which is used if less than n lines are drawn in Step 2, find the smallest uncovered number in the matrix. Add it to the elements in the covered rows and columns, and subtract it from all of the matrix elements. This step generates new zero elements, which provide additional possibilities for zero assignments.

As illustrated in the present example, the smallest uncovered number is 10. It is then necessary to add 10 to the fourth row and the first and fourth columns, and to subtract it from the entire matrix:

		after adding 10 to row 4			after adding 10 to columns 1 and 4		
0 (10) 20 30		0 10 20 30			10 10 20 40		
0 40 60 20		0 40 60 20			10 40 60 30		
0 20 20 0		0 20 20 0			10 20 20 10		
−20 −0 −0 −30 +10		30 10 10 40			40 10 10 50		
		+10			+10		

After subtracting 10 from the matrix it appears as

0	0	10	30
0	30	50	20
0	10	10	0
30	0	0	40

Step 3 can be shortened by subtracting 10 from the uncovered numbers and adding it to the numbers covered twice. The result is the same.

At the end of Step 3, Step 2 should be repeated. In Step 2 the minimal number of covering lines are drawn, which in this case is four.

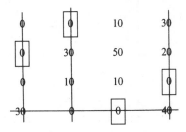

An optimal solution has now been reached. The solution of the assignment problem in terms of the original costs is

$$A2 + B1 + C4 + D3 = 60 + 10 + 20 + 10 = 100$$

Example of a Bidding Problem. Seven projects are to be completed, and seven different contractors are requested to submit bids. Each contractor is capable of performing any of the projects, but can complete only one. It may be further noted that the bid of contractor i on each project j is c_{ij}. The bids are submitted in thousands of dollars as

Contractor	Project						
	1	2	3	4	5	6	7
A	2	4	6	3	5	4	5
B	4	3	1	2	4	1	3
C	2	1	5	7	1	8	3
D	9	2	1	4	5	2	3
E	8	6	4	3	2	2	1
F	4	4	8	6	4	3	6
G	4	3	2	8	7	5	4

It is important to remember that each contractor can complete only one job. The objective function is to subcontract all the jobs at a minimal cost.

Step 1a
Subtract 2, 1, 1, 1, 1, 3, 2 from the respective rows.

Step 1b
Subtract 0, 0, 0, 1, 0, 0, 0 from the respective columns. The new cost matrix is

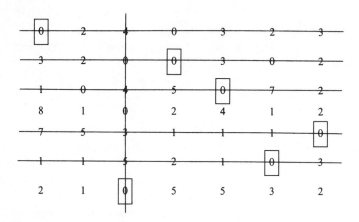

Step 2
To cover the zeros only six lines are used (to be doubly sure, check the six zero assignments that are available). Since seven lines are necessary to provide an optimal solution, proceed to Step 3.

Step 3
The smallest uncovered number is 1. Therefore, subtract 1 from all uncovered numbers and add 1 to all numbers covered twice. The new

cost matrix is

0	2	5	0	3	2	3
3	2	1	**0**	3	0	2
1	0	5	5	**0**	7	2
7	**0**	0	1	3	0	1
7	5	4	1	1	1	**0**
1	1	6	2	1	**0**	3
1	0	**0**	4	4	2	1

It is now possible to cover the zeros with seven lines, and an optimal solution has been reached. In terms of the original costs, the optimal solution is

$$A1 + B4 + C5 + D2 + E7 + F6 + G3$$
$$= 2 + 2 + 1 + 2 + 1 + 3 + 2 = 13 \quad \text{or} \quad \$13,000$$

Note that the total row subtractions in Step 1a are $2 + 1 + 1 + 1 + 1 + 3 + 2 = 11$, the column subtraction in Step 1b is 1, and the additional subtraction in Step 3 is 1, which results in a total of $11 + 1 + 1 = 13$. The value subtracted from the rows and columns is always equal to the optimal solution.

6.4. The Relationship Between Transportation and Assignment Problems

You have probably noticed a strong similarity between the classical transportation problem and the assignment problem. There is one major difference between these two types of problems. Instead of a general plant capacity s_i and demand d_j such as appear in the transportation problem, the capacities and demands of the assignment problem are always equal to 1. In short, the assignment problem is a special case of the transportation problem, since every time $s_i = 1$ for all i and $d_j = 1$ for all j these conditions constitute an assignment problem.

The transportation algorithm presented in Chapter 5 can also be applied to solve assignment problems. The number of basic variables required for a full basis is $n + n - 1$. In an assignment problem, however, only n assignments (positive variables) are possible. The solution is highly degenerate ($n - 1$ variables are at zero level and degenerate), and it may be difficult to determine the specific degenerate basic variables. Thus, even though it can be used, the transportation algorithm is not an efficient process for solving assignment problems.

While the assignment problem is a special case of the transportation problem, it is also true that the transportation problem can be viewed as a special case of the assignment problem. Consider the 2×3 transportation problem presented in Table 6.3. This problem can be easily formulated and solved as an assignment problem. Each source is divided into subsources, each of which has a capacity of 1. Thus, the first source is assigned to three subsources, s_1^1, s_1^2, s_1^3, and the second source is separated into two subsources s_2^1, s_2^2. Similarly, each destination is split into subdestinations, each of which has a demand of 1. More specifically, d_1 remains the same, d_2 is separated into d_2^1 and d_2^2, and d_3 is divided into d_3^1 and d_3^2. The costs associated with each subsource and subdestination remain the same as the original costs. The problem is now structured mathematically as

			Destination		
Source	$d_1 = 1$	$d_2^1 = 1$	$d_2^2 = 1$	$d_3^1 = 1$	$d_3^2 = 1$
$s_1^1 = 1$	4	5	5	7	7
$s_1^2 = 1$	4	5	5	7	7
$s_1^3 = 1$	4	5	5	7	7
$s_2^1 = 1$	1	6	6	7	7
$s_2^2 = 1$	1	6	6	7	7

The example can now be defined as an assignment problem, and the solution can be applied to the transportation model. We can then determine the optimal solution to the assignment problem as

$$s_1^1 d_2^1 = s_1^2 d_2^2 = s_1^3 d_3^2 = s_2^1 d_3^1 = s_2^2 d_1 = 1$$

Moreover, the solution to the transportation problem is

$$x_{12} = 2 \qquad x_{13} = 1 \qquad x_{21} = 1 \qquad x_{23} = 1$$

Since the transportation problem is a special case of the assignment problem, and the assignment problem is a special case of the transportation problem, the two are equivalent. (The proof is for when supply equals

Table 6.3. A Transportation Problem

		Destination	
Source	$d_1 = 1$	$d_2 = 2$	$d_3 = 2$
$s_1 = 3$	4	5	7
$s_2 = 2$	1	6	7

demand but this assumption can be relaxed through the use of dummy origins or destinations.)

Generally speaking, the algorithm for the assignment problem is easier to use and more efficient computationally than the one for the transportation problem. This characteristic might tempt you to convert all transportation problems into assignment problems. However, as the total capacity increases, this becomes impractical. If the total supply is large, for example, 100, the assignment problem is 100×100.

6.5. Further Treatment of the Assignment Problem

6.5.1. A Maximization Problem

So far, the criterion of performance in the illustrated cases has been to minimize total cost. An assignment problem may also have an objective function of maximizing profit. For instance, the measure of performance that is to be maximized can be defined as efficiency, profit, scores, utilization, or others. In order to provide a clear illustration, let us consider a fleet of five cars that must be sent to five customers. The profit in hundreds of dollars from the sale of each car to each customer is

Car	Customer				
	1	2	3	4	5
A	3	2	1	4	5
B	7	1	6	7	3
C	4	2	5	4	3
D	2	1	5	6	8
E	6	4	3	9	4

Rather than maximize the matrix, it is possible to convert the example into a minimization problem by making all the elements in the matrix negative (just as maximize $2x + 3y \Rightarrow$ minimize $-2x - 3y$). However, we can also add a constant to each row without changing the optimal assignment. So in order to eliminate the negative signs from the converted matrix, add the largest element in absolute value in the matrix to each matrix element; or, to obtain the same result, simply subtract the entire original matrix from its largest element. The case is then a minimization problem.

Car	Customer				
	1	2	3	4	5
A	6	7	8	5	4
B	2	8	3	2	6
C	5	7	4	5	6
D	7	8	4	3	1
E	3	5	6	0	5

Thus the procedure can now be applied to the new cost matrix. Step 1. After subtracting from each row and each column, the results are

2	[0]	4	1	0
[0]	3	1	0	4
1	0	[0]	1	2
6	4	3	2	[0]
3	2	6	[0]	5

which indicate an optimal solution of

$$A2 \quad B1 \quad C3 \quad D5 \quad E4$$

The solution in terms of the original profit is

$$2+7+5+8+9=31 \quad \text{or} \quad \$3100$$

6.5.2. Unequal Rows and Columns

An unbalanced problem is a situation where the number of rows does not equal the number of columns. For example, there are five regions to cover and only three salespersons are available, and again each salesperson can be assigned to only one region. The problem concerns determining which salespersons to assign to three of the regions. Two regions will not have sales representation. The profit table indicates the projected income in thousands of dollars if salesperson i is assigned to region j.

Salesperson	Region				
	1	2	3	4	5
Joe	3	2	1	4	3
Fred	5	4	2	1	4
Mary	1	3	4	5	2

Since the problem must be balanced, it is necessary to add two dummy salespersons. The profit generated by a dummy salesperson is always zero. Note that this profit can be any number, since the addition of a constant to a dummy row does not alter the decision problem (Property 1). After adding two dummy salespersons, the maximization problem is arranged as

	Region				
Salesperson	1	2	3	4	5
Joe	3	2	1	4	3
Fred	5	4	2	1	4
Mary	1	3	4	5	2
Dummy 1	0	0	0	0	0
Dummy 2	0	0	0	0	0

In order to convert the problem to a minimization problem follow the procedure of the last section and subtract the entire matrix from its largest entry which is a 5. Thus, the matrix may be expressed as

	Region				
Salesperson	1	2	3	4	5
Joe	2	3	4	1	2
Fred	0	1	3	4	1
Mary	4	2	1	0	3
Dummy 1	5	5	5	5	5
Dummy 2	5	5	5	5	5

After Step 2 the table appears as

Step 3 is to add and subtract 1 and yields

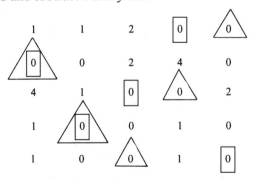

An optimal solution, enclosed in a box, is to assign Joe to region 4, Fred to region 1, Mary to region 3, and no one to regions 2 and 5. The total profit is

$$4+5+4=13 \quad \text{or} \quad \$13,000$$

An alternative solution, which is enclosed in triangles, is to assign Joe to region 5, Fred to region 1, Mary to region 4, and no one to regions 2 and 3. The total profit remains, as expected, $3+5+5=13$, or $13,000.

6.5.3. Impossible Assignments

In some cases it is impossible to assign job i to position j due to technological restrictions, unskilled labor, lack of training, contractor restrictions, or other constraints. The method applicable to this situation is basically the same as before except for one modification. Generally, wherever there is an impossible assignment, the cost associated with the constraints is extremely large. Consider the problem of completing four jobs on four machines. Job I cannot be completed on machine 2, and job III cannot be performed on machines 2 and 4. The time (in hours) for accomplishing each task on each machine is given as

	Machine			
Job	1	2	3	4
I	3	M	2	1
II	4	5	2	3
III	2	M	5	M
IV	4	3	2	1

The number M in this table is a large number, such as 100. It is now possible to solve the problem by the routine method. The table generated from the procedure (specifically, from Step 1) is

$$
\begin{array}{cccc}
2 & M & 1 & \boxed{0} \\
2 & 1 & \boxed{0} & 1 \\
\boxed{0} & M & 3 & M \\
3 & \boxed{0} & 1 & 0
\end{array}
$$

Notice that since M is a large number, the subtraction of 1 or 2 from M is insignificant and can be ignored during computations. Therefore, the optimal solution is

Job	Machine
I	4
II	3
III	1
IV	2

and the total time required for the four jobs is $1+2+2+3=8$ hours.

6.6. The Bottleneck Assignment Problem

In this section an interesting variation of the assignment problem is considered. The objective function is changed from

$$\text{minimize } \sum_{i=1}^{n} \sum_{j=1}^{n} c_{ij} x_{ij}$$

to minimize the maximum $c_{ij} x_{ij}$ over all i, j. The reason for this change becomes clear at the end of this example.

A foreman is taking a four-person work crew from Philadelphia to Atlantic City in order to fix equipment at Resorts National Hotel and Casino. The crew is leaving in one van and traveling together. The four workers are all capable of working on any machine, but due to different skills the length of time it takes each worker to fix each machine varies. The times (in minutes) are given in the accompanying table.

	Worker			
Device	Alan	Betsey	Carl	Doris
Roulette wheel (R)	80	120	125	140
Slot machine (M)	20	115	145	60
Change machine (C)	40	100	85	45
Security camera (S)	65	35	25	75

The crew will all return to Philadelphia together. Therefore they will leave Atlantic City when the last crew member finishes. Thus the foreman needs to assign the tasks so that the largest task time (rather than the sum of the task times) is minimized. The last crew member to finish creates a bottleneck, whence the name of the problem.

At first glance it appears that the problem cannot be solved by means of the assignment method. However, a two-step conversion process enables us to solve it using the Hungarian method. (You are asked to show that the process does indeed find the optimal solution.) The procedure is

Step 1. Rank the times in increasing order (ties may be broken arbitrarily).

Step 2. Substitute for each cell cost 2^{rank} and then use the Hungarian method.

In the example the first step yields the table of ranks given by

	Worker			
Device	A	B	C	D
R	9	13	14	15
M	1	12	16	6
C	4	11	10	5
S	7	3	2	8

The second step yields the table

	Worker			
Device	A	B	C	D
R	2^9	2^{13}	2^{14}	2^{15}
M	2^1	2^{12}	2^{16}	2^6
C	2^4	2^{11}	2^{10}	2^5
S	2^7	2^3	2^2	2^8

Thus, the problem to be solved is a standard assignment problem with the cost matrix given by

	Worker			
Device	A	B	C	D
R	512	8192	16384	32768
M	2	4096	65536	64
C	16	2048	1024	32
S	128	8	4	256

Using the Hungarian method, we obtain, after row and column subtraction, these tables:

0	7680	15872	32256
0	4094	65534	62
0	2032	1008	16
124	4	0	252

and

0	7676	15872	32240
0	4090	65534	46
0	2028	1008	0
124	0	0	236

The smallest uncovered number is 1008 which yields the following table.

0	6668	14864	32240
0	3082	64526	46
0	1020	0	0
1132	0	0	1244

After the appropriate addition and subtraction of 46, the last table is

0	6622	14818	32194
0	3036	64480	0
46	1020	0	0
1178	0	0	1244

The optimal solution has a bottleneck of maximum $\{80, 35, 85, 60\} = 85$.

6.7. Summary

The assignment problem is a special case of the general linear programming problem. It is highly degenerate since $n-1$ of the $2n-1$ basic variables are zero. However, its unique properties make it a simple problem to solve using the *Hungarian method*. It is seen that the assignment and transportation problems are, in fact, equivalent, but that again the assignment problem is degenerate when viewed as a transportation problem. A variation of the standard problem allows a change in optimization criteria and generates the *bottleneck assignment problem*.

References and Selected Readings

Cooper, L., and D. Steinberg 1974. *Methods and Applications of Linear Programming*. Philadelphia: Saunders.

Dantzig, G. B. 1963. *Linear Programming and Extensions*. Princeton, New Jersey: Princeton University Press.

Hesse, R., and G. Woolsey 1980. *Applied Management Science*. Chicago, Illinois: Science Research Associates.

Machol, R. E. 1976. *Elementary System Mathematics*. New York: McGraw-Hill.

Problems

1. Each of five temporary office workers is to be assigned to one of five different jobs. Each has ranked the jobs in order of preference, giving 1 to the most desirable job and 5 to the least desirable job. The rankings are given in the following table.

Worker	Job				
	Typing	Filing	Shorthand	Receptionist	Gopher
Ann	4	2	3	5	1
Betty	1	2	3	5	4
Cathy	1	4	5	2	3
Dick	5	Not qualified	4	2	1
Elmer	3	4	Not qualified	2	1

How should the assignment be made so that each office temporary gets the highest possible sum of ranks?

2. Mel's machine shop, with seven different machines, has received an order to do two jobs, each one requiring the use of three machines. The cost depends on both the machine and the job assigned, as follows.

	Machine						
Job	M_1	M_2	M_3	M_4	M_5	M_6	M_7
J_1	110	100	94	105	90	75	90
J_2	130	110	140	120	105	110	100

Mel's has a cost of $45 if any of the first four machines are left unassigned, and $20 for the last two. Find the minimal cost assignment.

3. Five jobs should be assigned to five workers. Find the minimal cost assignment for the cost matrix below.

	Workers				
Jobs	1	2	3	4	5
1	4	9	3	11	4
2	9	8	3	10	8
3	7	5	3	8	6
4	9	5	3	4	6
5	10	11	7	10	11

4. The Carzak Supply company has agreed to deliver four items as follows: one item at time t_1, one item at time t_2, one item at time t_3, and one item at time t_4 where $t_1 < t_2 < t_3 < t_4$. However, the items deteriorate with age, so the expected revenue from each item becomes less as the delivery date gets further in the future. Given the expected profits from each item at different dates, find the optimal delivery schedule.

	Time			
Item	t_1	t_2	t_3	t_4
1	$20	20	18	18
2	25	22	20	12
3	25	21	20	15
4	15	9	9	8

5. Solve the following assignment problem. (The entries represent cost.)

	1	2	3	4
1	2	4	2	4
2	8	5	4	5
3	4	6	8	9
4	8	4	2	4

6. Solve the following assignment problem where the entries represent profits.

	1	2	3	4	5
1	12	8	9	13	15
2	15	11	10	7	6
3	4	8	3	2	4
4	12	12	13	10	12
5	9	7	4	8	15

7. The Z and W Supply Company received eight orders to be fulfilled immediately. However, the manager has only five trucks available, and no truck can serve more than one order. Given the revenue matrix and the penalty cost for the incompleted orders, find the optimal delivery schedule.

	Orders							
Truck	1	2	3	4	5	6	7	8
1	70	35	48	75	65	65	56	40
2	50	46	32	×	20	50	55	52
3	38	×	45	56	55	48	80	25
4	15	62	18	×	20	35	50	×
5	55	25	24	50	42	46	22	30
Penalty cost	20	22	25	22	40	45	40	35

Note: A × indicates that this assignment is not possible

8. Hawkeye Manufacturing Company is introducing four new products. The company has five plants available to manufacture these products; due to constraints in capacity, however, only one product can be assigned to a single plant. The expected annual production costs at each plant for each product are given in the table. Find the assignments that minimize costs.

	Plants				
Product	1	2	3	4	5
A	$95,000	80,000	105,000	106,000	128,000
B	100,000	90,000	108,000	110,000	121,000
C	170,000	154,000	210,000	140,000	152,000
D	180,000	140,000	140,000	122,000	110,000

9. Solve the following assignment problem from its cost matrix.

	1	2	3	4	5
1	3	-1	4	6	2
2	6	4	2	5	1
3	1	1	4	2	0
4	5	2	3	1	4
5	5	2	4	2	6

10. Solve the following assignment problem, where the objective is a maximation.

	1	2	3	4	5
1	-2	8	4	2	7
2	4	2	8	4	8
3	5	6	10	5	4
4	8	4	12	10	12
5	6	3	5	4	5

11. Solve Problem 5 as a bottleneck assignment problem.

12. Solve Problem 4 as a bottleneck assignment problem.

13. Show that the bottleneck assignment method does indeed find the optimal solution.

Seven | Integer Programming

7.1. Introduction

The simplex procedure and its variants were developed and available for use in the early fifties. However, when this type of linear programming was applied to real-life problems, operations researchers soon found that it was not suitable for some cases. Specifically, many applications require that the decision variables be integer rather than continuous variables. For instance, if the decision variable is the number of airplanes or houses to build, there can be no fractional portion and the results must be integers (or whole numbers). The integer restriction appears to be a minor modification, but it can eventuate in a difficult problem. The first formulation of integer programming was presented in the early fifties and the first successful method for solving the integer programming problem was suggested by Gomory (1958) in 1956. Land and Doig (1960) suggested another approach for solving this problem, and many other solutions have also been introduced since then. In general, however, the methods for solving integer programming problems can be attributed either to Gomory or to Land and Doig. All the procedures for integer programming have a tendency to solve subclasses of problems but not the general integer programming problem itself. This matter remains unresolved, and excellent commercial packages for applications of large-scale integer programming are not yet available.

There are integer programming problems requiring that all the variables be integers; these are referred to as *pure integer programming* (PIP), or integer programming (IP). In other cases, called *mixed-integer programming* (MIP), some of the variables are restricted to being integers, whereas other variables are free to be fractional or integer.

When confronted with IP for the first time, your first impulse might be to ignore the integrality portion of the problem and to proceed to solve the problem as a linear programming problem, rounding off the solutions. This approach is a good one and may be applicable to many problems. For example, if the optimal linear programming solution is $x_1 = 1282.3$ and $x_2 = 345.1$, then the solution $x_1 = 1282$ and $x_2 = 345$ is probably not a bad procedure to follow for the integer programming problem. If the solution is a feasible one, then it is also very close to the optimal one (and most likely the optimal solution itself). However, there are many cases to which the round-off approach is not applicable. To appreciate this fact, consider the graphical problem given in Figure 7.1. The optimal linear programming solution is

$$x_1 = 2.9$$
$$x_2 = 3.3$$

It is apparent that the rounding off method is not successful, since the answer provides the following four possibilities:

$$
\begin{array}{ll}
x_1 = 3 & x_2 = 3 \\
x_1 = 3 & x_2 = 4 \\
x_1 = 2 & x_2 = 3 \\
x_1 = 2 & x_2 = 4
\end{array}
$$

Figure 7.1. Integer Programming—Graphical Example

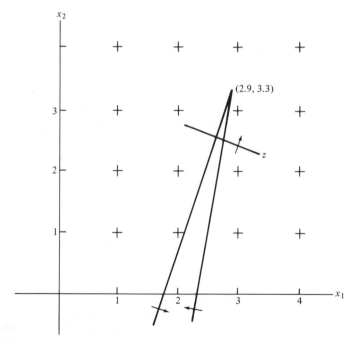

All four solutions are infeasible and furthermore far from the optimal solution $x_1 = 2$, $x_2 = 0$ (which happens also to be the only feasible solution). Note that integer points are also called *lattice points*. This example illustrates a case where rounding off is inappropriate in seeking a solution.

There are basically three approaches for solving integer programming problems. The first can be defined as a cutting plane approach. At each iteration, new constraints are added to reduce the feasible region until an integer solution is reached. The second approach is a branch and bound method, in which at each iteration the problem is split into two subproblems. This is similar to a decision tree process. Each subproblem provides an upper or lower bound for the optimal solution. The third approach is to completely enumerate all the solutions. Some of the solutions are explicitly evaluated while others are implicitly considered. This procedure was developed by Balas (1965) for solving 0–1 programming problems (where the variables are restricted to the values of zero or one).

7.2. Gomory's Cut

Gomory's basic cutting plane idea is to disregard the integer restrictions and to solve the remaining problem as a linear programming problem. The process ends if the linear programming solution is an integer one, and he suggests adding a new constraint if it is a fractional one. This new constraint has the following properties (the proof can be found in Hu, 1970):

1. It eliminates the optimal linear programming solution from the feasible region.
2. It reduces the area of the feasible region.
3. It does not eliminate any of the integer feasible solutions. (All lattice points remain as before.)
4. It passes through at least one integer point.

Once the new constraint is added to the linear programming problem, the modified problem is solved without regard for the integer restrictions. Moreover, the technique is based on solving a sequence of linear programming problems. Eventually an integer solution is reached if one exists. These steps can be summarized in the following chart:

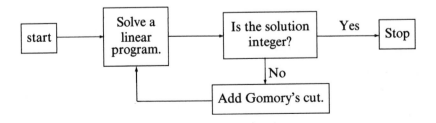

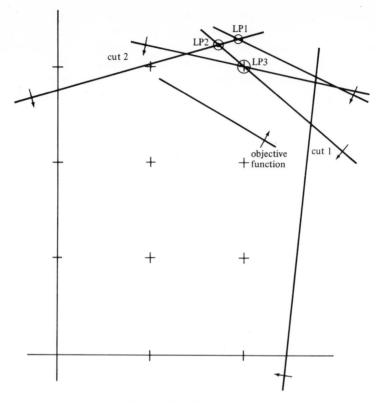

Figure 7.2. Gomory's Cut

A graphical illustration of Gomory's cut appears in Figure 7.2. The original linear programming problem results in an optimal solution at LP1. Cut 1 is then added and the new linear programming problem is optimized. The second linear programming solution is at LP2. Then the second cut is added to the linear programming problem and the new problem is solved. The third linear programming solution is at LP3, which is an integer point. Therefore, LP3 is also the integer programming solution to the original problem. It is important to point out that both cuts have reduced the area of the feasible region; pass through an integer point (the optimal one), and have not eliminated any integer feasible solution.

7.2.1. The Development of Gomory's Cut

Any number a_j can be divided into its integer part, denoted by $[a_j]$, and the fractional part, denoted by f_j (greater than or equal to zero, but strictly less than one). For example,

	Integer part	Fractional part
$5.3 = 5 + 0.3$	5	0.3
$4.999 = 4 + 0.999$	4	0.999
$-5.6 = -6 + 0.4$	-6	0.4
$7 = 7 + 0$	7	0
$-2 = -2 + 0$	-2	0
$-2.01 = -3 + 0.99$	-3	0.99
$a_j = [a_j] + f_j$	$[a_j]$	f_j
	where $[a_j]$ is integer	$0 \leqslant f_j < 1$

Gomory's cut is based on the property that if there is a constraint

$$a_1 x_1 + a_2 x_2 + \cdots + a_n x_n = a_0 \qquad (7.1)$$

where $x_j \geqslant 0$ and integer for all j, then the following constraint must also hold true:

$$f_1 x_1 + f_2 x_2 + \cdots + f_n x_n \geqslant f_0 \qquad (7.2)$$

where $f_1, f_2, \ldots, f_n, f_0$ are the fractional parts of $a_1, a_2, \ldots, a_n, a_0$, respectively. Equation (7.2) is called Gomory's cut and forms the conceptual framework for the integer programming algorithm. It is first important to pause and reconsider some facts before providing easy generalizations.

To demonstrate that (7.1) implies (7.2), substitute $a_j = [a_j] + f_j$ into (7.1) and thus derive

$$([a_1] + f_1) x_1 + ([a_2] + f_2) x_2 + \cdots + ([a_n] + f_n) x_n = [a_0] + f_0$$

or

$$[a_1] x_1 + [a_2] x_2 + \cdots + [a_n] x_n - [a_0] = f_0 - f_1 x_1 - f_2 x_2 - \cdots - f_n x_n$$

The left-hand side is an integer number; thus the right-hand side is also an integer number. The right-hand side is smaller than or equal to f_0 since $0 \leqslant f_j < 1$ and $x_j \geqslant 0$ for all j. Also, since $f_0 < 1$ then

$$f_0 - f_1 x_1 - f_2 x_2 - \cdots - f_n x_n \leqslant 0$$

because the left-hand side is an integer number and smaller than or equal to f_0, which is strictly smaller than 1; therefore, it must be smaller than or equal to 0.

7.2.2. An Example of Gomory's Cut

Gomory's cut, given by Equation (7.2), may be used in an algorithmic way to solve integer programming problems. To demonstrate its versatility, let us now solve the following problem.

The Orland Construction Company is engaged in building two types of homes, a medium-priced home on a 0.25-acre lot and an expensive one on a 0.4-acre lot. The Orland Company has recently purchased 3 acres on which to build homes; however, zoning restrictions limit the number of medium-priced homes to 8 and the more expensive homes to 4. Furthermore, the Orland Company has estimated that a medium-priced home results in a $10,000 profit, and an expensive home nets a $20,000 profit. Thus, the problem may be expressed as: Determine the number of medium-priced and expensive homes that the Orland Company should build to maximize its profit.

Let x be the number of medium-priced homes and y the number of expensive homes. The objective function is

$$\text{maximize profit } P = 10x + 20y \text{ in thousands of dollars}$$

The constraints are

$$0.25x + 0.4y \leqslant 3 \quad \text{the acreage restriction}$$
$$x \leqslant 8$$
$$y \leqslant 4$$
$$x, y \geqslant 0 \text{ and integer}$$

After multiplying the acreage constraint by 20 (to avoid the fractional coefficients) the problem is

$$\text{Maximize} \qquad\qquad P = 10x + 20y$$
$$\text{subject to} \qquad 5x + 8y \leqslant 60$$
$$x \leqslant 8$$
$$y \leqslant 4$$
$$x, y \geqslant 0 \text{ and integer}$$

The first step is to ignore the integer restrictions and proceed to solve the problem as a linear programming problem.

$$\text{Maximize} \quad P = 10x + 20y + 0s_1 + 0s_2 + 0s_3$$
$$\text{subject to} \quad 5x + 8y + s_1 \qquad\qquad = 60$$
$$x \qquad\qquad + s_2 \qquad = 8$$
$$y \qquad\qquad + s_3 = 4$$
$$x, y, s_1, s_2, s_3 > 0$$

Table 7.1. Orland Construction Company Linear Programming Solution

c_b	Basic variables	Value	10 x	20 y	0 s_1	0 s_2	0 s_3
0	s_1	60	5	8	1	0	0
0	s_2	8	1	0	0	1	0
0	s_3	4	0	①	0	0	1
		0	−10	−20	0	0	0
0	s_1	28	⑤	0	1	0	−8
0	s_2	8	1	0	0	1	0
20	y	4	0	1	0	0	1
		80	−10	0	0	0	20
10	x	5.6	1	0	0.2	0	−1.6
0	s_2	2.4	0	0	−0.2	1	1.6
20	y	4	0	1	0	0	1
		136	0	0	2	0	4

Notice that the slack variables can be written as

$$s_1 = 60 - 5x - 8y \tag{7.3}$$

$$s_2 = 8 - x \tag{7.4}$$

$$s_3 = 4 - y \tag{7.5}$$

The solution to the linear programming problem is presented in Table 7.1. It is reached after two iterations and is given by

$$x = 5.6$$
$$y = 4$$
$$s_2 = 2.4$$
$$\text{profit} = 136$$

The solution is also presented in Figure 7.3 in a graphical form, and the optimal solution is indicated as (5.6, 4).

In this example the solution is not an integer one and the first Gomory's cut is now generated. From the final tableau of Table 7.1 we could use either the first or the second row as the generating row for the Gomory's cut because both $x = 5.6$ and $s_2 = 2.4$ are nonintegers.

In this case choose the first because the fractional part of 5.6 is 0.6, and the fractional part of 2.4 is 0.4. Generally, it is better to take the one with the largest fractional part—it reflects a "deeper" cut of the feasible region. Thus, the source row from which the first Gomory's cut is generated is

$$1 \cdot x + 0 \cdot y + 0.2 \cdot s_1 + 0 \cdot s_2 - 1.6 \cdot s_3 = 5.6$$

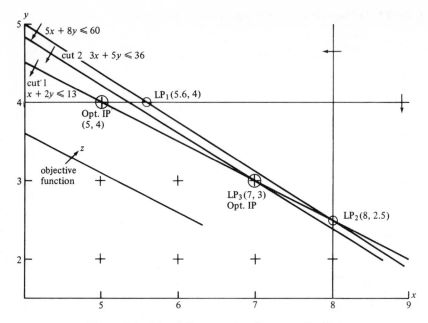

Figure 7.3. Orland Construction Company Problem

The fractional parts of the coefficients are

0 for x
0 for y
0.2 for s_1
0 for s_2
0.4 for s_3 $(-1.6 = -2 + 0.4)$
0.6 for the free element which is the same as the right-hand side.

Therefore, the first Gomory cut is

$$0.2s_1 + 0.4s_3 \geq 0.6$$

(Note that to plot the cut, s_1 and s_3 must have coordinates in terms of x and y.) After multiplication by -1 and addition of the slack variable $s_4 \geq 0$, the constraint is

$$-0.2s_1 - 0.4s_3 + s_4 = -0.6$$

or

$$s_4 = 0.2s_1 + 0.4s_3 - 0.6 \qquad (7.6)$$

Consequently, this cut is now added to the optimal tableau of Table 7.1.

The problem consisting of the original three constraints, the new constraint just generated, the original variables (x, y, s_1, s_2, s_3), and the new variable s_4 is given in Table 7.2. The current solution, $x = 5.6$, $s_2 = 2.4$, $y = 4$,

Table 7.2. Orland Construction Company—Intermediate Table

c_b	Basic variables	Value	10 x	20 y	0 s_1	0 s_2	0 s_3	0 s_4
10	x	5.6	1	0	0.2	0	−1.6	0
0	s_2	2.4	0	0	−0.2	1	1.6	0
20	y	4	0	1	0	0	1	0
0	s_4	−0.6	0	0	−0.2	0	(−0.4)	1
		136	0	0	2	0	4	0
10	x	8	1	0	1	0	0	−4
0	s_2	0	0	0	−1	1	0	4
20	y	(2.5)	0	1	−0.5	0	0	2.5
0	s_3	1.5	0	0	0.5	0	1	−2.5
		130	0	0	0	0	0	10

$s_4 = -0.6$, and profit $= 136$, is not feasible because $s_4 < 0$, even though it satisfies the optimality criteria. Notice that all the elements in the last row are nonnegative, thus satisfying the optimality condition.

Since the solution is not feasible, it can be handled by the dual simplex method presented in Section 4.2. Recall that the dual simplex method is easily applicable when an infeasible solution satisfies the optimality criteria. The algorithm maintains the optimality criteria and brings the infeasible solution to a feasible one. The process has two steps and the first is to determine the variable that leaves the basis. This is s_4 because the infeasibility is caused by $s_4 = -0.6$. The second step is to determine the name of the variable that replaces s_4. This is accomplished by taking the following ratio: Divide all elements of the last row by the corresponding elements of the row for s_4 providing the denominator is negative. The entering variable is the one with the largest ratio.

$$\text{Maximum} \left\{ \frac{2}{-0.2}, \frac{4}{-0.4} \right\} = -10$$

There is a tie between variables s_1 and s_3, so we arbitrarily choose s_3 as the entering variable to replace s_4. The pivot element is now -0.4 (always negative) and we continue to perform the regular pivot operation. The result is given by the second tableau of Table 7.2. The solution is an optimal (degenerate) one expressed as

$$x = 8 \quad s_2 = 0 \quad y = 2.5 \quad s_3 = 1.5 \quad \text{profit} = 130$$

This solution is marked LP$_2$ in Figure 7.3. (The plotting of Gomory's cut is discussed in the following pages.)

The solution is optimal for the linear programming problem but not for the integer programming problem. The source row for the second Gomory's cut can be derived either from $y = 2.5$ or from $s_3 = 1.5$. Arbitrarily, we take the former. The source row is then

$$0x + 1y - 0.5s_1 + 0s_2 + 0s_3 + 2.5s_4 = 2.5$$

and the fractional coefficients are

0 for x
0 for y
0.5 for s_1 $(-0.5 = -1 + 0.5)$
0 for s_2
0 for s_3
0.5 for s_4
0.5 for the free element

which is identical to the right-hand side. Thus, the second Gomory's cut is

$$0.5s_1 + 0.5s_4 \geqslant 0.5$$

After multiplying by -1 and adding a new slack variable $s_5 \geqslant 0$, we have as the constraint

$$-0.5s_1 - 0.5s_4 + s_5 = -0.5$$

which is added to the optimal tableau of Table 7.2, and presented in Table 7.3. The solution for the problem given in Table 7.3 is infeasible, but satisfies the optimality criteria. The dual simplex method is applied to the problem. The variable that leaves the basis is s_5 (since $s_5 = -0.5 < 0$) and it is replaced by s_1 since maximize$\{0/-0.5, 10/-0.5\} = 0$ occurs at s_1. A pivot operation is performed around the pivot element -0.5 that results in an optimal solution defined as

$$x = 7 \qquad s_2 = 1 \qquad y = 3 \qquad s_3 = 1 \qquad s_1 = 1 \qquad \text{profit} = 130$$

It is emphasized that this is an integer solution, and indeed, the optimal solution to the original integer programming problem. More specifically, in Figure 7.3 it is defined as LP$_3$ and also as Opt. IP.

The optimal integer programming solution is not unique, since the coefficient in the last row for the nonbasic variable s_5 is zero, and an alternative optimal solution for the linear programming problem exists. In an effort to determine the other solution, force s_5 into the basis, and the variable that leaves the basis is s_3 (since minimize $\{7/2, 1/1\} = 1/1$ occurs at s_3). Also, the pivot element is 1, and after a pivot operation around the number, the alternative optimal linear programming solution integer solution is

$$x = 5 \qquad s_2 = 3 \qquad y = 4 \qquad s_5 = 1 \qquad s_1 = 3 \qquad \text{profit} = 130$$

Note that the fact that the coefficient of s_5 is zero in the last tableau guarantees that an alternative linear programming solution exists but does not ensure that an alternative integer programming solution exists. It just happens that the last tableau is also an integer solution.

Table 7.3. Orland Construction Company—Final Tableau

c_b	Basic variables	Value	x	y	s_1	s_2	s_3	s_4	s_5
10	x	8	1	0	1	0	0	-4	0
0	s_2	0	0	0	-1	1	0	4	0
20	y	2.5	0	1	-0.5	0	0	2.5	0
0	s_3	1.5	0	0	0.5	0	1	-2.5	0
0	s_5	-0.5	0	0	$\boxed{-0.5}$	0	0	-0.5	1
		130	0	0	0	0	0	10	0
10	x	7	1	0	0	0	0	-5	2
0	s_2	1	0	0	0	1	0	5	-2
20	y	3	0	1	0	0	0	3	-1
0	s_3	1	0	0	0	0	1	-3	$\boxed{1}$
0	s_1	1	0	0	1	0	0	1	-2
Optimal LP		130	0	0	0	0	0	10	0
10	x	5	1	0	0	0	-2	1	0
0	s_2	3	0	0	0	1	2	-1	0
20	y	4	0	1	0	0	1	0	0
0	s_5	1	0	0	0	0	1	-3	1
0	s_1	3	0	0	1	0	2	-5	0
Alternative IP solution		130	0	0	0	0	0	10	0

In addition, the two Gomory's cuts can be plotted on a two-dimensional graph and the first cut is

$$0.2s_1 + 0.4s_3 \geqslant 0.6$$

Substitute s_1 from (7.3) and s_3 from (7.5) and thus derive

$$0.2(60 - 5x - 8y) + 0.4(4 - y) \geqslant 0.6$$

or

$$12 - x - 1.6y + 1.6 - 0.4y \geqslant 0.6$$

or

$$x + 2y \leqslant 13$$

This defines cut 1 and it is plotted in Figure 7.3.

The second Gomory's cut is

$$0.5s_1 + 0.5s_4 \geqslant 0.5$$

Substitute s_4 from (7.6) to obtain

$$0.5s_1 + 0.5(0.2s_1 + 0.4s_3 - 0.6) \geqslant 0.5$$

or

$$0.5s_1 + 0.1s_1 + 0.2s_3 - 0.3 \geqslant 0.5$$

or

$$0.6s_1 + 0.2s_3 \geqslant 0.8$$

and substitute s_1 from (7.3) and s_3 from (7.5) to obtain

$$0.6(60 - 5x - 8y) + 0.2(4 - y) \geqslant 0.8$$

or

$$36 - 3x - 4.8y + 0.8 - 0.2y \geqslant 0.8$$

or

$$3x + 5y \leqslant 36$$

This is plotted as cut 2 in Figure 7.3. Notice that both cuts pass through an integer point $x=7$, $y=3$ and neither one eliminates any integer feasible solutions from the feasible region.

The optimal solution for the Orland Construction Company is to build $x=7$ medium-priced and $y=3$ expensive homes, or alternatively to build $x=5$ medium-priced and $y=4$ expensive homes. In any event, the profit is $130,000.

Notice that the optimal solution $x=5$, $y=4$ is a boundary point, since $y=4$ is located on the constraint $y\leqslant 4$. However, the solution $x=7$, $y=3$ is an interior point and is not situated on any of the boundaries. Moreover, $x=7$, $y=3$ certainly is below the zoning restriction of $x\leqslant 8$, $y\leqslant 4$ and the acreage requirement is

$$0.25 \cdot 7 + 0.4 \cdot 3 = 2.95$$

less than the 3 acres that are available. An interior point is never an optimal solution in linear programming, but an interior point may be optimal with integer programming problems. We emphasize that Gomory's cut, as presented in this section, is for all integer programming problems. For mixed-integer programming problems, a modification of Gomory's cut is necessary.

7.3. Branch and Bound Method

The concept of branch and bound was first introduced by Land and Doig (1960). The basic idea is to divide the feasible region into smaller regions. This is simply a procedure that assigns lower and upper bounds to the solution while systematically solving the subproblems. For example, suppose that x_5 is designated to be integer. We can then solve the problem using linear programming and part of the solution is $x_5 = 3.4$. Then, a branching process occurs with either $x_5 \leqslant 3$ or $x_5 \geqslant 4$ materializing into two subproblems. Consequently, the solution of the two subproblems provides bounds (either upper or lower depending on whether it is a maximization or minimization problem) for the original problem. This problem demonstrates the important advantages of this method, such as implementing a sequence of linear programming algorithms to solve a nonlinear programming problem. This method is applicable to all integer problems, mixed-integer problems, and some nonlinear programming problems (either the constraints or the objective function may be nonlinear).

In order to utilize the branch and bound method, the only requirements are the repeated solving of linear programming problems with some "bookkeeping" for the upper and lower bounds.

The key point of the procedure is the partitioning of the feasible region. Therefore, the method is very efficient for bounded variables. For the sake of discussion, suppose $2 \leqslant x_5 \leqslant 6$. Then partition $x_5 = 2$, 3, 4, 5, or 6. Moreover, the method is more efficient if there are more restrictions and bounds on the variables.

7.3.1. A Maximization Example

Looking back over the previous example, let us now solve the Orland Construction Company problem presented in Section 7.2. The problem is stated as

$$\text{Maximize} \qquad P = 10x + 20y$$

$$\text{subject to} \qquad 5x + 8y \leqslant 60$$

$$x \leqslant 8$$

$$y \leqslant 4$$

$$x, y \geqslant 0 \text{ and integer}$$

First, ignore the integer restriction and solve the problem as a linear programming problem (call the problem P_1) to obtain the solution

$$x = 5.6 \qquad y = 4 \qquad P = 136$$

The profit $P = 136$ is an upper bound for the problem simply because any additional restriction of integrality can only reduce the solution from $P = 136$.

While $y = 4$ is an integer, the value for $x = 5.6$ violates the integrality restriction. A branching process is now performed by forming two new problems P_2 and P_3 for $x \leqslant 5$ and $x \geqslant 6$ as follows:

P_2: Maximize $P = 10x + 20y$ P_3: Maximize $P = 10x + 20y$

  subject to $5x + 8y \leqslant 60$ subject to $5x + 8y \leqslant 60$

$x \leqslant 8$ $x \leqslant 8$

$y \leqslant 4$ $y \leqslant 4$

$x \leqslant 5$ $x \geqslant 6$

$x, y > 0$ $x, y \geqslant 0$

Also, notice that in P_2, $x \leqslant 8$ is redundant to $x \leqslant 5$. The two problems P_2 and P_3 are now solved as LP problems. The solution for P_2 is

$$x=5 \qquad y=4 \qquad P=130$$

and for P_3

$$x=6 \qquad y=3.75 \qquad P=135$$

In effect, the solution for P_2 is integer and is feasible for the original problem. It should be emphasized that the value $P=130$ is a lower bound for the original problem and it is possible that other IP solutions may exist in the range 130–136.

The solution pertaining to P_3 now provides us with a new upper bound of $P=135$. As indicated before, the optimal solution is at least 130 and at most 135.

Since $y=3.75$ in P_3, a new branching process is used to determine a solution by using either $y \leqslant 3$ or $y \geqslant 4$. Consequently, two new problems are structured, such as P_4 and P_5:

P_4:   Maximize $10x+20y$ P_5:   Maximize $10x+20y$

subject to $5x+8y \leqslant 60$ subject to $5x+8y \leqslant 60$

$$x \leqslant 8 \qquad\qquad\qquad\qquad\qquad x \leqslant 8$$

$$y \leqslant 4 \qquad\qquad\qquad\qquad\qquad y \leqslant 4$$

$$x \geqslant 6 \qquad\qquad\qquad\qquad\qquad x \geqslant 6$$

$$y \geqslant 4 \qquad\qquad\qquad\qquad\qquad y \leqslant 3$$

$$x, y \geqslant 0 \qquad\qquad\qquad\qquad\quad x, y \geqslant 0$$

Notice that whenever a branching is performed and new problems are formulated, all the restrictions imposed on the problems are maintained. For example, $x \geqslant 6$ was imposed on P_3, and is therefore maintained as a constraint when P_3 is partitioned into P_4 and P_5. Here, however, other factors appear to surface that are related to the problem, such as that the solution for P_4 is infeasible and no further consideration is given to the branches beyond P_4, since any new restrictions cannot change the infeasibility condition.

The solution to P_5 is

$$x=7.2 \qquad y=3 \qquad P=132$$

Therefore, the new upper bound has been reduced from 135 to 132 and the optimal solution is somewhere between 130 and 132. Since $x=7.2$ in P_5, a new branching step is required. It is taken with respect to x (since $y=3$ is an integer). Moreover, P_5 is now partitioned into P_6 and P_7 with either $x \leqslant 7$ or

$x \geqslant 8$ and presented as follows:

P_6:	Maximize $10x + 20y$	P_7:	Maximize $10x + 20y$
	subject to $5x + 8y \leqslant 60$		subject to $5x + 8y \leqslant 60$
	$x \leqslant 8$		$x \leqslant 8$
	$y \leqslant 4$		$y \leqslant 4$
	$x \geqslant 6$		$x \geqslant 6$
	$y \leqslant 3$		$y \leqslant 3$
	$x \leqslant 7$		$x \geqslant 8$
	$x, y \geqslant 0$		$x, y \geqslant 0$

The list of constraints of P_6 can be stated as

$$5x + 8y \leqslant 60 \qquad 6 \leqslant x \leqslant 7 \qquad y \leqslant 3 \qquad x, y \geqslant 0$$

and for P_7 the constraints are

$$5x + 8y \leqslant 60 \qquad x = 8 \qquad y \leqslant 3 \qquad x, y \geqslant 0$$

The solution to P_7 is

$$x = 8 \qquad y = 2.5 \qquad P = 130$$

and the solution for P_6 is

$$x = 7 \qquad y = 3 \qquad P = 130$$

The new upper bound has been reduced from 132 to 130, which means that the lower bound is 130 and the upper bound is 130, and consequently the optimal solution must be 130. Obviously, we already have a solution with a value of 130 that is determined in P_2. Thus $x = 5$, $y = 4$, $P = 130$ of P_2 is an optimal solution. However, in this particular case, it just happened that another optimal solution is found in P_6, which is

$$x = 7 \qquad y = 3 \qquad P = 130$$

Any further partitioning of P_7 with $y \leqslant 2$ or $y \geqslant 3$ can provide us with integer solutions that are worse than P_7 or lower than $P = 130$, which is inferior to what already exist.

The branch and bound solution is summarized in Figures 7.4 and 7.5, which demonstrate the various steps of the algorithm, the six linear programming subproblems (seven LP problems), and feasible regions and optimal solutions.

7.3.2. A Minimization Example

The branch and bound method for minimization problems is very similar to the maximization problem solved in Section 7.3.1. Let us now examine the

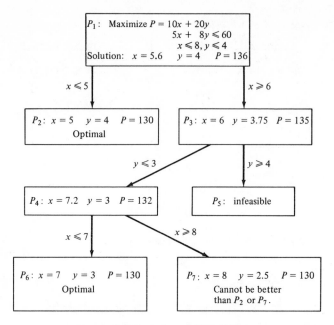

Figure 7.4. Branch and Bound Tree

following problem to demonstrate the similarity between the different approaches.

$$\text{Minimize} \qquad c = 2x + 3y$$

$$\text{subject to} \qquad x + 3y \geq 5$$

$$2x + y \geq 6$$

$$x, y \geq 0 \text{ and integer}$$

The problem is first solved as a linear programming problem called P_1 and the solution is

$$x = 2.6 \qquad y = 0.8 \qquad c = 7.6$$

Also, the graphical solution of the problem is illustrated in Figure 7.6. The value $c = 7.6$ is a lower bound for the integer programming problem.

The first branch is around the integer x, which cannot be $x = 2.6$ but must be either $x \leq 2$ or $x \geq 3$. Thus, problems P_2 and P_3 are constructed as

P_2: Minimize $c = 2x + 3y$ P_3: Minimize $c = 2x + 3y$

      subject to $x + 3y \geq 5$ subject to $x + 3y \geq 5$

                $2x + y \geq 6$ $2x + y \geq 6$

                    $x \leq 2$ $x \geq 3$

                  $x, y \geq 0$ $x, y \geq 0$

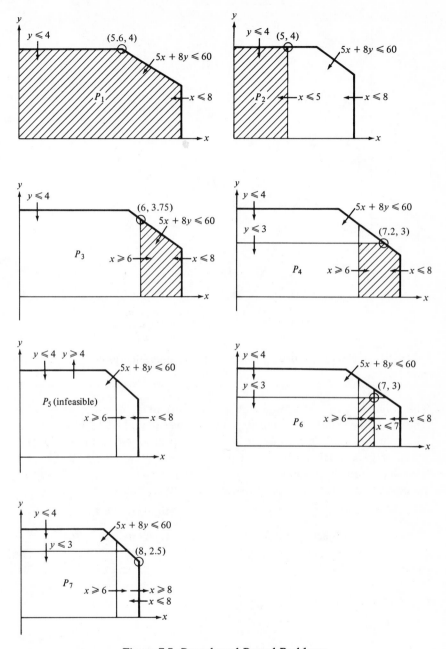

Figure 7.5. Branch and Bound Problems

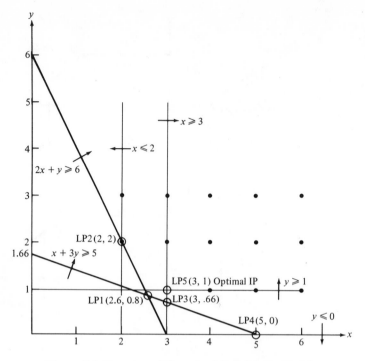

Figure 7.6. Branch and Bound—Minimization Problem

The solution to P_2 is $x=2$, $y=2$, $c=10$, and to P_3 is $x=3$, $y=\frac{2}{3}$, $c=8$. Therefore, the value $c=8$ from P_3 is a new lower bound that replaces $c=7.6$. The solution to P_2 is an integer with an upper bound to the problem of $c=10$. The optimal solution is at best $c=8$ and at worst $c=10$. The next branch is from P_3 around $y=\frac{2}{3}$ with either $y\leqslant0$ or $y\geqslant1$. This branching process generates the two problems depicted by P_4 and P_5 as

P_4: Minimize $c=2x+3y$ P_5: Minimize $c=2x+3y$

$\quad$ subject to $x+3y\geqslant5$ $\qquad\qquad$ subject to $x+3y\geqslant5$

$\qquad\qquad\qquad 2x+y\geqslant6$ $\qquad\qquad\qquad\qquad 2x+y\geqslant6$

$\qquad\qquad\qquad x\geqslant3$ $\qquad\qquad\qquad\qquad\quad x\geqslant3$

$\qquad\qquad\qquad y\leqslant0$ $\qquad\qquad\qquad\qquad\quad y\geqslant1$

$\qquad\qquad\qquad x,y\geqslant0$ $\qquad\qquad\qquad\qquad\quad x,y\geqslant0$

The solution to P_4 is $x=5$, $y=0$, $c=10$ and to P_5 it is $x=3$, $y=1$, $c=9$. Moreover, the solution to the linear programming problem P_5 is $c=9$, which is a new lower bound. However, the solution to P_5 is also an integer, and therefore it is also an upper bound. Since $c=9$ is both an upper and lower

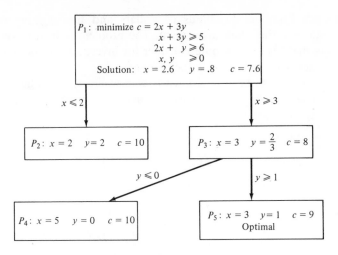

Figure 7.7. Branching Process

bound, then $x=3$, $y=1$, $c=9$ is also the optimal solution. In addition, P_4 also happens to be an integer solution but its value $c=10$ is worse than $c=9$.

Thus, the example may be expressed as the branching process summarized in Figure 7.7.

7.4. Balas Implicit Enumeration Technique

A special kind of integer programming occurs where the variables can obtain only two values, 0 or 1. The problem is then called a 0–1 programming problem. The 0–1 restriction can represent an on–off situation, such as to buy or not to buy or to invest or not to invest. An efficient algorithm called *implicit enumeration* was developed by Balas (1965) for the 0–1 programming problem. Consider the following capital budgeting problem. The Gail-Car Investment Company is examining the possibilities of five different projects for future investment opportunities. The company has decided to invest no more than $10,000 in each of the first two years, and $8000 in each of the next two years. The five projects, the investments, and the returns of each one are given as

	Project				
Investment in	1	2	3	4	5
year 1	2	4	0	3	2
year 2	2	1	5	3	-2
year 3	3	-2	4	4	2
year 4	3	3	5	0	2
Net return after expense	14	17	15	11	14

The data are listed in thousands of dollars and the negative numbers indicate returns rather than investment. The problem is then to determine the projects that Gail-Car should consider for investment without violating the restrictions while maximizing their return.

Let y_j, $j=1,2,\ldots,5$, be a decision variable where $y_j=0$ indicates not to choose project j and $y_j=1$ means to invest in project j. The integer programming problem is now

$$\text{Maximize} \qquad 14y_1 + 17y_2 + 15y_3 + 11y_4 + 14y_5$$

$$\text{subject to} \qquad 2y_1 + 4y_2 + 0y_3 + 3y_4 + 2y_5 \leqslant 10$$

$$2y_1 + y_2 + 5y_3 + 3y_4 - 2y_5 \leqslant 10$$

$$3y_1 - 2y_2 + 4y_3 + 4y_4 + 2y_5 \leqslant 8$$

$$3y_1 + 3y_2 + 5y_3 + 0y_4 + 2y_5 \leqslant 8$$

$$y_j = 0,1 \qquad j=1,2,\ldots,5$$

The problem must be brought to a standard form before applying the conventional implicit enumeration procedure. The following conditions must be satisfied in the standard form:

Condition 1. The objective function has to be a minimization one. This can be achieved simply by minimizing $\Sigma_j(-c_jy_j)$ rather than maximizing $\Sigma_j c_j y_j$.

The new objective function is then

$$\text{Minimize} \ -14y_1 - 17y_2 - 15y_3 - 11y_4 - 14y_5$$

Condition 2. The coefficient of the variables in the objective function must be positive. In order to achieve this, make the following transformation of variables

$$y_j = 1 - x_j$$

for each negative objective function coefficient. The substitution of $y_j = 1-x_j$ is for both the objective function and the constraints. This change results in

$$\text{Minimize} \quad -71 + 14x_1 + 17x_2 + 15x_3 + 11x_4 + 14x_5$$

$$\text{subject to} \qquad 11 - 2x_1 - 4x_2 + 0x_3 - 3x_4 - 2x_5 \leqslant 10$$

$$9 - 2x_1 - x_2 - 5x_3 - 3x_4 + 2x_5 \leqslant 10$$

$$11 - 3x_1 + 2x_2 - 4x_3 - 4x_4 - 2x_5 \leqslant 8$$

$$13 - 3x_1 - 3x_2 - 5x_3 + 0x_4 - 2x_5 \leqslant 8$$

$$x_j = 0,1 \qquad j=1,2,\ldots,5$$

Condition 3. The constraints must be "greater than or equal to" the right hand side. This is easily accomplished by moving the elements to the right-hand side. Notice that the constant -71 in the objective function does not affect the decision process and can be ignored. The problem is now in its standard form and may be expressed as

$$\text{Minimize} \qquad 14x_1 + 17x_2 + 15x_3 + 11x_4 + 14x_5$$

$$\text{subject to } g_1: \qquad -1 + 2x_1 + 4x_2 + 0x_3 + 3x_4 + 2x_5 \geq 0$$

$$g_2: \qquad 1 + 2x_1 + x_2 + 5x_3 + 3x_4 - 2x_5 \geq 0$$

$$g_3: \qquad -3 + 3x_1 - 2x_2 + 4x_3 + 4x_4 + 2x_5 \geq 0$$

$$g_4: \qquad -5 + 3x_1 + 3x_2 + 5x_3 + 0x_4 + 2x_5 \geq 0$$

$$x_j = 0, 1 \qquad j = 1, 2, \ldots, 5$$

where g_i refers to the ith constraint.

In working through the solution process one or more of the variables is "frozen" at level 0 or at level 1. Suppose that at iteration k, $x_3 = 1$, $x_2 = 0$ and x_1, x_4, x_5 are free variables. For notational purposes, the partial solution is denoted by $S_k = (3, -2)$. S_k contains the subscripts of the frozen variables where a positive subscript indicates that the variable is frozen at level 1 and a negative subscript, at level 0. If a partial solution is $S_k = (-3, -2, 4, 1)$, then $x_3 = 0$, $x_2 = 0$, $x_4 = 1$, $x_1 = 1$, and x_5 is free. $S_k = \emptyset$ means that no variable is restricted.

In the current problem, since there are five variables and each one can attain two values, there are $2^5 = 32$ different solutions. The implicit enumeration procedure searches the 32 solutions, with some solutions calculated explicitly and some implicitly. If a set of solutions is infeasible or inferior to an existing solution, this set is discarded and need not be calculated to find the optimal solution. The goal is to evaluate all 32 solutions in a systematic order either explicitly or implicitly. Figure 7.8 indicates the order in which computations are done. The figure resembles a tree with branches extending downward. At the top, $S_0 = \emptyset$ indicates an empty partial solution where no variables are restricted. Moreover, there are six levels in the figure. At the top level no variable is restricted, while at each succeeding level one additional variable is restricted.

The next step in the general process is to start with the smallest possible solution value, $S_0 = \emptyset$. You are advised to look simultaneously at Figure 7.8 and Table 7.4. Notice that since no variables are restricted, choosing all variables equal to zero minimizes the objective function. The values for the constraints are $g_1 = -1, g_2 = 1, g_3 = -3, g_4 = -5$, or $(-1, 1, -3, -5)$, which is infeasible. In order to get closer to feasibility it is necessary to raise $x_1 = 1$. In effect, this is an arbitrary decision since we could start with x_2, x_3, x_4, or

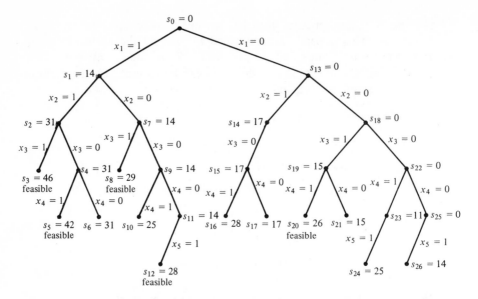

Figure 7.8. Balas Implicit Enumeration Technique

x_5. The only consideration here is to choose a variable that has the potential of assisting in satisfying the violated constraints. Note that constraints 1, 3, and 4 are violated when $S_0 = \varnothing$, and the coefficients of x_1 are 2, 3, and 3 in these rows. However, $S_1 = 1$ is still not feasible, since the constraints are $(1, 3, 0, -2)$ and the fourth constraint is violated. Moreover, the value of the objective function increases to 14. The next step is to raise $x_2 = 1$ to obtain a partial solution $S_2 = (1, 2)$, which satisfies the fourth constraint but violates the third. The operation itself is termed a *forward step*. The constraint values are $(5, 4, -2, 1)$ and the objective function is $14 + 17 = 31$. The search for a feasible solution succeeds in the next partial solution $S_3 = (1, 2, 3)$ with constraint values $(5, 9, 2, 6)$. This is a feasible solution with the objective function value of 46. It is apparent that other partial solutions descending from $S_3 = (1, 2, 3)$ are inferior to 46. The reason is simply that the descending solution adds positive values to the solution, because the coefficients of all variables in the objective function are positive.

Since it is not necessary to do a forward step, continue to proceed up the tree and test any branches that have not been examined. This operation is known as "backtracking." Beginning at $S_3 = (1, 2, 3)$, proceed to $S_2 = (1, 2)$, and branch to $S_4 = (1, 2, -3)$. The partial solution is infeasible, since $g_3 = -2$ and by adding $x_4 = 1$ to the partial solution, the result is $S_5 = (1, 2, -3, 4)$. In this instance, the value 42 is better than the previous 46 and S_5 is also a feasible solution. Thus, there is no reason to search among its descendants, which must have higher objective function value. A backtracking operation is now accomplished from S_5 to S_4, and a branch to $S_6 = (1, 2, -3, -4)$ is

Table 7.4

Solution	Constraints (g_1, g_2, g_3, g_4)	Objective function value
$s_0 = 0$	$-1, 1, -3, -5$	0
$s_1 = 1$	$1, 3, 0, -2$	14
$s_2 = 1, 2$	$5, 4, -2, 1$	31
$s_3 = 1, 2, 3$	$5, 9, 2, 6$ feasible	46
$s_4 = 1, 2, -3$	$5, 4, -2, 1$	31
$s_5 = 1, 2, -3, 4$	$8, 7, 2, 1$ feasible	42
$s_6 = 1, 2, -3, -4$	$5, 4, -2, 1$	31
$s_7 = 1, -2$	$1, 3, 0, -2$	14
$s_8 = 1, -2, 3$	$1, 8, 4, 3$ feasible	29
$s_9 = 1, -2, -3$	$1, 3, 0, -2$	14
$s_{10} = 1, -2, -3, 4$	$4, 6, 4, -2$	25
$s_{11} = 1, -2, -3, -4$	$6, 4, 6, 0$	39
$s_{12} = 1, -2, -3, -4, 5$	$3, 1, 2, 0$ feasible	28
$s_{13} = -1$	$-1, 1, -3, -5$	0
$s_{14} = -1, 2$	$3, 2, -5, -2$	17
$s_{15} = -1, 2, -3$	$3, 2, -5, -2$	17
$s_{16} = -1, 2, -3, 4$	$6, 5, -1, -2$	28
$s_{17} = -1, 2, -3, -4$	$3, 2, -5, -2$	17
$s_{18} = -1, -2$	$-1, 1, -3, -5$	0
$s_{19} = -1, -2, 3$	$-1, 6, 1, 0$	15
$s_{20} = -1, -2, 3, 4$	$2, 9, 5, 0$ feasible	26
$s_{21} = -1, -2, 3, -4$	$-1, 6, 1, 0$	15
$s_{22} = -1, -2, -3$	$-1, 1, -3, -5$	0
$s_{23} = -1, -2, -3, 4$	$2, 4, 1, -5$	11
$s_{24} = -1, -2, -3, 4, 5$	$4, 2, 3, -3$	23
$s_{25} = -1, -2, -3, -4$	$-1, 1, -3, -5$	0
$s_{26} = -1, -2, -3, -4, 5$	$1, -1, -1, -3$	14

still infeasible. Normally, a forward step should be done, but here the next solution $(1, 2, -3, -4, 5)$ raises the value of the objective function from 31 to 45. Although this is feasible, it is inferior to the previous value of 42. Therefore, forward steps are no longer necessary. A backtracking step from S_6 to S_1 is suggested on the "unchecked branch." The partial solution now is $S_7 = (1, -2)$, which is infeasible since $g_4 = -2$. Adding $x_3 = 1$ to S_7 results in $S_8 = (1, -2, 3)$, which is a feasible solution and has an objective function value of 29. No descendants exist with a better solution, and a backtracking operation to S_7 and branching to $S_9 = (1, -2, -3)$ is performed. The solution is infeasible and a forward step to $S_{10} = (1, -2, -3, 4)$ is still infeasible. The constraint values are $(4, 6, 4, -2)$ and the objective function value is 25. Another forward step to $(1, -2, -3, 4, 5)$ results in an objective function value higher than the 29 found so far. A backtrack is now in order, to S_9, and the next solution $S_{12} = (1, -2, -3, -4, 5)$ is feasible and furthermore yields only the value 28. Now backtrack to S_{11}, branch to

$(1, -2, -3, -4, -5)$, which you will find infeasible, and proceed upward toward S_0 (the first node with only one branch checked). Then branch to $S_{13} = (-1)$, which is infeasible; adding $x_2 = 1$ to the solution results in $S_{14} = (-1,2)$, which is still infeasible and has a value of 17 for the objective function. By adding $x_3 = 1$, we cause the objective function to exceed the current value of 28 by 4 units. Therefore, we go down the tree to $S_{15} = (-1,2,-3)$ with constraint values of $(3,2,-5,-2)$, which is infeasible. Adding $x_4 = 1$, we obtain $S_{16} = (-1,2,-3,4)$, which is still infeasible and has a value of 28 for the objective function. Since any further branching could make the objective function worse than this solution value of 28, return to S_{15} and branch to $S_{17} = (-1,2,-3,-4)$. This solution is still infeasible and its descendants are inferior to the objective function value of 28, so we backtrack to S_{13} and go to

$$S_{18} = (-1,-2) \qquad S_{19} = (-1,-2,3) \qquad S_{20} = (-1,-2,3,4)$$

Finally, S_{20} is a feasible solution with the objective function value of 26 and constraint values of $(2,9,5,0)$. It can be shown that the other remaining solutions in the tree are either infeasible or inferior to S_{20}. It is apparent that the optimal solution has been reached.

The solution found for this example is

$$x_1 = 0 \qquad x_2 = 0 \qquad x_3 = 1 \qquad x_4 = 1 \qquad x_5 = 0$$

The original decision variables are $y_j = 1 - x_j$. Therefore, the final solution is $y_1 = 1$, $y_2 = 1$, $y_3 = 0$, $y_4 = 0$, $y_5 = 1$ and the profit is $14 + 17 + 14 = 45$, or \$45,000.

Improving the Implicit Enumeration Method. Even though the problem has 32 distinct solutions, the foregoing procedure requires the evaluation of 26 combinations of the variables for feasibility and objective function value. This does not seem to be an efficient example. However, the algorithm can be improved with a better choice of the sequence of variables. The modification is minor; as before, any infeasible partial solution must be augmented with a new variable. The purpose of the new variable is to assist in bringing the solution to feasibility. For example, previously when the solution was $S_0 = \emptyset$, $x_1 = 1$ was added. The coefficients for the activity matrix in column x_1 are $2+2+3+3=10$, while the coefficients for x_2 are $4+1-2+3=6$; $0+5+4+5=14$ for x_3; $3+3+4+0=10$ for x_4; $2-2+2+2=4$ for x_5. The modification is that the variables are ranked according to descending sum order, which results in 14, 10, 10, 6, 4 for x_3, x_1, x_4, x_2, x_5, respectively. This ranking indicates that x_3 enters before x_1 and x_1 before x_2, etc. This modification improves the algorithm considerably.

This procedure is summarized in Table 7.5. Begin with $S_0 = \emptyset$ and with the constraints 1, 3, and 4 violated. Among the variables that can bring the solution to a feasible one, choose x_3. At $S_1 = (3)$, constraint 1 is violated and from among variables x_1, x_2, x_4, and x_5, choose x_1 to enter at level 1 (it is a

Table 7.5

Step	Solution S	Violated constraints V	Objective function limit	Variables for consideration to enter solution T	Constraints that cannot be satisfied	Variable to enter S	Upper bound value
0	∅	1,3,4		1,2,3,4,5		3	
1	3	1		1,2,4,5		1	
2	3,1	∅					29
3	3,−1	1	14	4			
4	3,−1,4	∅					26
5	3,−1,−4	1	11	∅			
6	−3	1,3,4	26	1,2,4,5		1	
7	−3,1	4	12		4		
8	−3,−1	1,3,4	26	2,4,5		4	
9	−3,−1,4	4	15	5	4		
10	−3,−1,−4	1,3,4	26	2,5	3		

tie between the variables x_1 and x_4, and we arbitrarily take x_1). Then $S_2 = (3,1)$ is a feasible solution with value 29. The next step is $S_3 = (3,-1)$, which has a value of 15 for the objective function. The difference $(29-15 = 14)$ means that no variable enters the partial solution if its contribution to the objective function is 14 or more. Since x_4 is such a variable, the set $S_4 = (3,-1,4)$ is feasible with a value 26.

A new column—"constraints that cannot be satisfied"—appears in Table 7.5. It is possible for a partial solution such as $S_7 = (-3,1)$ to have a constraint that can never be satisfied. For $x_3 = 0$, $x_1 = 1$ and an objective function limit of 12, there is no way to satisfy the fourth constraint. The only two variables that have positive coefficients in the fourth constraint are x_2 and x_5, but the objective function coefficients are higher than 12. Also, for $S_{10} = (-3,-1,-4)$, the third constraint cannot be satisfied, since the only positive variable is x_5 with a coefficient of 2, which cannot satisfy -3. In this modified approach, only 10 partial solutions are tested, as compared to the 26 used by the previous method.

In the implicit enumeration technique and its modified version some solutions are compiled explicitly and some are computed implicitly. The effectiveness of the technique is due to the number of implicit computations. Obviously, our goal is to calculate as many solutions implicitly as possible. Here are two cases that increase the efficiency of the implicit enumeration solutions:

1. No solutions descending from a feasible partial solution should be calculated. Their objective function value is inferior to a feasible one.
2. All solutions descending from a feasible partial solution should not be considered if either there is a constraint that can never be satisfied,

or the value of the objective function of the partial solution is greater than an existing feasible solution.

The solution found in this example is

$$x_1 = 0 \quad x_2 = 0 \quad x_3 = 1 \quad x_4 = 1 \quad x_5 = 0$$

The original decision variables are y and $y_j = 1 - x_j$. Therefore, the final solution is $y_1 = 1$, $y_2 = 1$, $y_3 = 0$, $y_4 = 0$, $y_5 = 1$ and the profit is $14 + 17 + 14 = 45$, or \$45,000.

7.5. The Use of Dummy 0–1 Variables

Many mathematical programming problems can be simplified and converted into integer linear programming problems by the use of dummy 0–1 variables. A dummy 0–1 variable is an integer variable that attains the values 0 or 1. It is important to realize that a mathematical programming problem that is complicated in formulation and solution can be made relatively easy by being translated into an integer programming problem with 0–1 variables. The following are some examples in which dummy 0–1 variables can be used.

7.5.1. Either–Or Constraints

In this model, a mathematical programming problem may have "either–or" types of constraints. Some examples may involve a decision to send a truck to a customer either with at least a 500 lb load or not at all. In another situation, a generator can be set to work effectively either at up to 5000 rpm or at least 8000 rpm. Still another case may involve the cost of traveling from New York City to Philadelphia which is \$12 if traveling by bus or \$16 if traveling by train. Mathematically these three examples are

$$\text{either } x \geqslant 500 \text{ lb or } x = 0 \qquad \text{(truck load)}$$

$$\text{either } x \leqslant 5000 \text{ or } x \geqslant 8000 \qquad \text{(generator speed)}$$

$$\text{either } x = 12 \text{ or } x = 16 \qquad \text{(trip cost)}$$

This structured constraint can be easily handled by a dummy 0–1 variable, δ.

Suppose a problem is

$$\begin{aligned}
\text{Maximize} \qquad & 2x + 3y \\
\text{subject to} \qquad & x + y \leqslant 5 \\
& x, y \geqslant 0 \\
\text{and either} \qquad & x \leqslant 1 \text{ or } x \geqslant 2
\end{aligned}$$

Now add the variable $\delta=0,1$ and solve the problem with μ being a large number, say $\mu=100$.

$$\text{Maximize } 2x+3y+0\cdot\delta$$

$$\text{subject to } x+y\leqslant5$$

$$x\leqslant1+\delta\cdot\mu$$

$$x\geqslant2-(1-\delta)\mu$$

$$x, y\geqslant0 \qquad \delta=0,1$$

This is a *mixed-integer programming problem* (MIP) in which some of the variables, such as x and y, are continuous, and one of the variables, δ, is an integer. The two cases for $\delta=0$ and $\delta=1$ are

	for $\delta=0$		for $\delta=1$
Maximize	$2x+3y$	Maximize	$2x+3y$
subject to	$x+y\leqslant5$	subject to	$x+y\leqslant5$
	$x\leqslant1$		$x\leqslant1+\mu$
	$x\geqslant2-\mu$		$x\geqslant2$
	$x, y\geqslant0$		$x, y\geqslant0$

For $\delta=0$ the constraint $x\leqslant1$ appears, while $x\geqslant2-\mu$ is redundant for $\mu=100$. For $\delta=1$ the constraint $x\leqslant1+\mu$ is redundant for $\mu=100$ and $x\geqslant2$ appears. The program itself decides whether $\delta=0$ or $\delta=1$ should be taken, according to the objective function maximize $2x+3y$.

7.5.2. K Out of N Constraints

A company may have N constraints such that of these N constraints at least K must be satisfied. Let the N constraints be

$$g_i\leqslant b_i \qquad i=1,2,\ldots,N$$

Let us now introduce N new variables

$$\delta_i=0,1 \qquad i=1,2,\ldots,N$$

and μ is a large number.

The constraint set is now written as

$$g_i\leqslant b_i+(1-\delta_i)\mu$$

$$\sum_i\delta_i\geqslant K$$

For each $\delta_i=0$, $g_i\leqslant b_i+(1-0)\mu$ becomes redundant for large μ and for each $\delta_i=1$ the constraint is $g_i\leqslant b_i+(1-1)\mu=b_i$, which is the original constraint. The restriction $\Sigma\delta_i \geqslant K$ guarantees that at least K of the δ_i are 1, which means that constraint i is an active constraint. If we want to have exactly K active constraints out of N, this is accomplished by replacing $\Sigma\delta_i \geqslant K$ by $\Sigma\delta_i = K$. Consider the following example:

$$\text{Maximize} \qquad 2x+3y$$
$$\text{subject to} \qquad x+y \leqslant 5$$
$$2x+y \leqslant 4$$
$$x, y \geqslant 0$$

At least one of the constraints $x \leqslant 4$, $y \leqslant 4$, $x+y \leqslant 9$ must be true.
   Introduce δ_1, δ_2, δ_3 and $\mu = 100$ and formulate the problem as

$$\text{Maximize} \qquad 2x+3y$$
$$\text{subject to} \qquad x+y \leqslant 5$$
$$2x+y \leqslant 4$$
$$x \leqslant 4+(1-\delta_1)100$$
$$y \leqslant 4+(1-\delta_2)100$$
$$x+y \leqslant 9+(1-\delta_3)100$$
$$\delta_1+\delta_2+\delta_3 \geqslant 1$$
$$x, y \geqslant 0 \qquad \delta_1, \delta_2, \delta_3 = 0, 1$$

If δ_1 is 1, then $x \leqslant 4$ is active. If δ_2 is 1, the second constraint is active. If $\delta_3 = 1$, then $x+y \leqslant 9$ is active.

7.5.3. Multidimensional Either–Or Constraints

This example demonstrates a linear objective function and a feasible region that consists of mutually exclusive regions. The problem cannot be formulated as a linear programming problem because of the either–or constraints (nonconvex feasible region). With the dummy variables, however, the problem can be formulated as a mixed-integer programming problem. For example, let the feasible region be the shaded area in Figure 7.9 and let the objective function be maximize $2x+3y$.
   The feasible region is one of the following three regions:

Either	$0 \leqslant x \leqslant 3$	$0 \leqslant y \leqslant 2$
or	$5 \leqslant x \leqslant 7$	$0 \leqslant y \leqslant 3$
or	$2.9 \leqslant x \leqslant 5.3$	$3 \leqslant y \leqslant 5$.

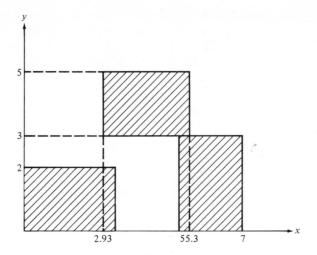

Figure 7.9.

Add $\delta_1, \delta_2, \delta_3$, and a large number μ and formulate the list of constraints:

$$0 \leqslant x \leqslant 3 + (1-\delta_1)\mu$$
$$0 \leqslant y \leqslant 2 + (1-\delta_1)\mu$$
$$5 - (1-\delta_2)\mu \leqslant x \leqslant 7 + (1-\delta_2)\mu$$
$$0 \leqslant y \leqslant 3 + (1-\delta_2)\mu$$
$$2.9 - (1-\delta_2)\mu \leqslant x \leqslant 5.3 + (1-\delta_3)\mu$$
$$3 - (1-\delta_2)\mu \leqslant y \leqslant 5 + (1-\delta_3)\mu$$
$$\delta_1 + \delta_2 + \delta_3 = 1$$
$$x, y \geqslant 0 \qquad \delta_1, \delta_2, \delta_3 = 0, 1$$

with the objective function maximize $2x + 3y$. Only one δ_j is equal to 1 and this is the active region. For $\delta_j = 0$ the jth region becomes redundant.

7.5.4. The Fixed Charge Problem

Consider an example of the cost of renting a car. Assume that there is a fixed charge of \$18 a day plus 25¢ per mile for renting the car. If the mileage is zero, however, then there is no need to pay the fixed charge. Thus

$$\text{cost} = \begin{cases} 18 + 0.25x & \text{if} \quad x > 0 \\ 0 & \text{if} \quad x = 0 \end{cases}$$

The objective function is

$$\text{Minimize} \qquad 18\delta + 0.25x$$

$$\text{subject to} \qquad \delta \cdot \mu \geqslant x$$

$$\delta = 0, 1$$

If x is positive, then δ must be 1 (μ is a large number) and a charge of $\delta \cdot 18 = 18$ is made in the objective function. When $x = 0$, however, there is no need for δ to be positive, and in a minimization δ stays at zero.

There are many other problems that can be simplified and converted into integer programming. The main idea is similar to the ones presented here; unfortunately, the scope of this book does not permit more examples.

7.6. Summary

In this chapter the variables in the basic linear programming problem have been restricted to being integers only, and this restriction increases the variety of situations that can be formulated as mathematical programming problems. This is evident from the last section of the chapter, which presents either–or types of constraints, K out of N options, and on–off conditions such as the fixed charge problem.

Many integer programming problems have a finite number of solutions, but this number is generally large enough to preclude explicit enumeration of all possible solutions. Thus a decision maker must evaluate the solutions by means of an *implicit enumeration* scheme or a *branch and bound* method. Alternatively, problems may be solved using the *cutting plane* idea.

In the next chapter some integer programming problems will be solved efficiently by means of dynamic programming. Also, an analysis of problems that are nonlinear is begun.

References and Selected Readings

Abadie, J., ed. 1970. *Integer and Nonlinear Programming*. New York: American Elsevier.

Balas, E. 1965. An additive algorithm for solving linear programs with zero–one variables. *Oper. Res.* 13: 517–546.

Garfinkel, R., and G. Nemhauser 1972. *Integer Programming*. New York: John Wiley and Sons, Inc.

Gomory, R. E. 1958. An algorithm for the integer solutions to linear programs. *Princeton IBM Math. Res. Report* November. Also in *Recent Advances in Mathematical Programming*, ed. R. L. Graves and P. Wolfe, pp. 269–302. New York: McGraw-Hill, 1963.

Hu, T. C. 1970 *Integer Programming and Network Flows*. Reading, Massachusetts: Addison-Wesley.

Land, A. H., and A. G. Doig 1960. An automatic method for solving discrete programming problems. *Econometrica* 28 (3): 497–520.

Plane, D. R., and C. McMillan, Jr. 1971. *Discrete Optimization.* Englewood Cliffs, New Jersey: Prentice-Hall, Inc.

Salkin, H. M. 1975. *Integer Programming.* Reading Massachusetts: Addison-Wesley.

Zionts, S. 1974. *Linear and Integer Programming.* Englewood Cliffs, New Jersey: Prentice-Hall, Inc.

Problems

1. Bert Vance is a production manager who has four machines available for the production of 400 units of product. The setup cost, cost per unit, and capacity in units of each machine is shown in the accompanying table. How should Bert schedule the production so that total costs are minimized?

Machine	Cost per unit	Capacity	Setup cost
A	10	800	100
B	4	1400	200
C	2	1000	300
D	5	1500	200

2. Tom Fix has to do ten different jobs during a period of four hours. Each job takes different times to perform, and jobs cannot be performed simultaneously. Tom receives a profit for each completed job. These profits together with the times it takes to do each job are shown in the accompanying table. How should Tom schedule his activities in order to maximize profits?

	1	2	3	4	5	6	7	8	9	10
Time[a]	24	35	40	15	45	30	20	55	35	10
Profits	$200	$250	$300	$220	$350	$250	$200	$355	$310	$180

[a] In minutes.

3. Minimize $3x_1 + 5x_2$

   subject to $2x_1 - 3x_2 \geqslant -8$

   $16x_1 + 8x_2 \geqslant 50$

   $x_1, x_2 \geqslant 0$ and integer

4. Minimize $4x_1 + 6x_2 + 5x_3$

   subject to $8x_1 - 5x_2 + 9x_3 \geqslant 10$

   $x_1 + x_2 - 2x_3 \geqslant 8$

   $x_1, x_2, x_3 \geqslant 0$ and integer.

5. Maximize $5x_1 + 6x_2 + 7x_3$

   subject to $9x_1 + 12x_2 + 5x_3 \leqslant 40$

   $6x_1 + 4x_2 + 20x_3 \leqslant 30$

   $x_1, x_2, x_3 \leqslant 0$ and integer

6. Maximize $x_1 + x_2$

   subject to $2x_1 + 5x_2 \leqslant 16$

   $6x_1 + 5x_2 \leqslant 30$

   $x_1, x_2 \geqslant 0$ and integer.

7. A job shop has received an order which calls for the manufacturing of ten units of each of three types of parts. There are five machines capable of producing any of the parts. The cost per unit of producing each part at each machine is given in the accompanying table. Formulate the problem as an integer programming problem.

	Machine				
Part	1	2	3	4	5
1	10	12	14	20	15
2	9	8	8	10	7
3	4	5	7	3	9

8. Minimize $4x_1 + 6x_2 + 5x_3$

   subject to $x_1 - 4x_2 + 2x_3 \leqslant 10$

   $8x_1 + x_2 \leqslant 8$

   $-2x_1 + 3x_2 \leqslant 12$

   $x_1, x_2, x_3 \geqslant 0$ and integer.

9. Maximize $x_1 + 3x_2 + 10x_3$

   subject to $x_1 + 4x_2 \leqslant 7$

   $x_2 + 3x_3 \leqslant 8$

   $3x_1 + x_2 + x_3 \leqslant 17$

   $x_1, x_2, x_3 \geqslant 0$ and integer

10. Bimbo's, a restaurant chain, has identified five potential sites for new establishments. It also has a total of $200,000 available for immediate use. The costs as well as the expected revenue at each potential site are given in the table below. Determine the optimal number of new restaurants and their sites.

	Site				
	1	2	3	4	5
Cost	$40,000	$35,000	$60,000	$75,000	$40,000
Revenue	$60,000	$55,000	$66,000	$80,000	$60,000

11. Solve Problem 8 of Chapter 2, assuming that fractional units are not permitted.

Eight | Dynamic Programming

8.1. Introduction Through an Example

The preceding chapters have focused on mathematical programming problems with linear constraints and objectives. In this chapter we treat the analysis of specially structured problems involving nonlinear functions. The general methods for solving these nonlinear programming problems are discussed in Chapter 9. The type of problem considered here is one that can be treated as a sequence of smaller, easier to solve subproblems. In fact, this conversion of one large problem into a sequence of smaller problems is the essence of dynamic programming. Typically dynamic programming is a useful method for solving problems with integer variables and/or combinatorial problems. Initially, problems with linear functions (with integer variables) are considered; then the concepts are extended to nonlinear functions.

The concept of decomposing a large problem into a sequence of smaller ones has existed for quite some time. However, the unifying concepts of dynamic programming, and in fact the name "dynamic programming" itself, are due to Bellman (Bellman and Dreyfus, 1962). The technique is even more recent than linear programming, since the first book on dynamic programming appeared in 1957. Dynamic programming is widely used for production planning, network problems, resource allocation problems, and the control of engineering processes.

In previous chapters, it was possible to define the model by a set of constraints and an objective function. Thus the optimal procedure for solving the problem was developed for a general model. The characteristics of dynamic programming are slightly different. Dynamic programming is an approach to solving wide variety of problems that in general are less structured than those encountered previously. To demonstrate as many types of problems as possible, we present as many numerical examples as possible. Therefore, to introduce the nature of dynamic programming, we consider first a resource allocation problem. In Section 8.2, a generalized framework for solving dynamic programming problems is presented, and finally in Section 8.3 several other numerical examples are presented.

Capital Budgeting Example. A state government has received proposals for five projects. Table 8.1 gives the cost of each project and its estimated value. The state budget for new projects is $15,000, and a legislative subcommittee must decide which of these projects should be approved and funded. Of course, the committee would like to maximize the total value that accrues due to project funding.

In mathematical programming terms the problem is expressed as

$$\text{Maximize} \qquad 8x_1 + 6x_2 + 2x_3 + 7x_4 + 5x_5$$

$$\text{subject to} \qquad 4x_1 + 6x_2 + 5x_3 + 3x_4 + 7x_5 \leqslant 15$$

$$x_j = \begin{cases} 1 & \text{if project } j \text{ is funded} \\ 0 & \text{otherwise} \end{cases} \qquad j = 1,2,3,4,5$$

This integer programming problem is known as a capital budgeting problem.

In general, a problem is a capital budgeting problem if it is of the form:

$$\text{Maximize} \ \sum_{j=1}^{n} v_j x_j$$

$$\text{subject to} \ \sum_{j=1}^{n} c_j x_j \leqslant b$$

$$x_j = 0 \text{ or } 1 \qquad j = 1,2,\dots,n$$

Notice that if this were a linear programming problem, there would be only one basic variable. Also, without the budget constraint there are 2^n possible

Table 8.1. Capital Budgeting Problem

Project (j)	Cost[a] (c_j)	Value (v_j)
1	4	8
2	6	6
3	5	2
4	3	7
5	7	5

[a] In thousands of dollars.

Table 8.2. Complete Enumeration of All Solutions to the Capital Budgeting Problem

Projects funded	Cost	Value
none	0	0
1	4	8
2	6	6
3	5	2
4	3	7
5	7	5
1, 2	10	14
1, 3	9	10
1, 4	7	15
1, 5	11	13
2, 3	11	8
2, 4	9	13
2, 5	13	11
3, 4	8	9
3, 5	12	7
4, 5	10	12
1, 2, 3	15	16
1, 2, 4	13	21[a]
1, 2, 5[b]	17	
1, 3, 4	12	17
1, 3, 5[b]	16	
1, 4, 5	14	20
2, 3, 4	14	15
2, 3, 5[b]	18	
2, 4, 5[b]	16	
3, 4, 5	15	14
1, 2, 3, 4[b]	18	
1, 2, 3, 5[b]	22	
1, 2, 4, 5[b]	20	
1, 3, 4, 5[b]	19	
2, 3, 4, 5[b]	21	
1, 2, 3, 4, 5[b]	25	

[a] This is the optimal value.
[b] This is an infeasible solution.

solutions to the problem, since each of the projects is either funded or not funded. Thus, a small problem like this five-project example could be solved by identifying all solutions, eliminating those that are not feasible, and choosing the best of the feasible solutions, as presented in Table 8.2. Of the feasible solutions the best choice is to fund projects 1, 2, and 4 which would yield a value of 21 at a cost of $13,000. (Notice that only $13,000 of the $15,000 are used, for lack of a better combination.) Obviously, this complete enumeration approach is not useful when a larger number of projects must be evaluated; a different method is needed for larger problems.

For this capital budgeting problem let $P(j, b)$ denote the optimal value of a problem with projects $j, j+1, \ldots, n$ and budget b. In other words $P(1,15)$ is the answer to the problem, $P(3,7)$ is the optimal value for a problem with projects 3, 4, and 5 and a budget of 7, and so forth.

There are two possibilities: Project 1 is either funded or not funded. If project 1 is funded, then there remain $15-4=\$11{,}000$ for projects 2 through 5, and this $\$11{,}000$ should be allocated in the best possible way among the last four projects. That is, if project 1 is funded, it requires the solution of $P(2,11)$. The total value would then be 8 for project 1 plus $P(2,11)$. If project 1 is not funded, then there are $\$15{,}000$ left for funding projects 2 through 5 and this should be done in an optimal fashion; that is, find $P(2,15)$. The total value in this case is 0 for project 1 plus $P(2,15)$. It follows then that the best value for the five-project problem is given by

$$P(1,15) = \text{maximum}\{8 + P(2,11), 0 + P(2,15)\} \qquad (8.1)$$

Notice that inside the braces there are two similar terms, $P(2,11)$ and $P(2,15)$. In order to solve the original problem we must solve a four-project problem for two different budgets, $\$11{,}000$ and $\$15{,}000$. Once $P(2,11)$ and $P(2,15)$ are known, it is simple to determine $P(1,15)$ according to Equation (8.1). Thus, solving the four-project problem enables us to solve the five-project problem.

Equation (8.1) can be rewritten as

$$P(1,15) = \max_{\substack{x_1=0,1 \\ 4x_1 \leqslant 15}} \{8x_1 + P(2, 15 - 4x_1)\}$$

Now let $b_2 = 15 - c_1 x_1$. That is, b_2 is the budget after project 1 is or is not funded. We know that b_2 must be either $\$15{,}000$ or $\$11{,}000$ but we consider b_2 to be any budget between 0 and $\$15{,}000$, which implies that the search is for $P(2, b_2)$. As in the case of Equation (8.1), project 2 is either funded or not. Hence

$$P(2, b_2) = \max_{\substack{x_2=0,1 \\ 6x_2 \leqslant b_2}} \{6x_2 + P(3, b_2 - 6x_2)\}$$

That is, if there are not sufficient funds for the second project $(c_2 > b_2)$, then x_2 must be zero. Otherwise, either do or do not fund project 2, depending on whether or not $6 + P(3, b_2 - 6)$ is larger than $0 + P(3, b_2)$. Again, given $P(3, b_2 - 6)$ and $P(3, h_2)$ the four-project problem is trivial because it depends only on two three-project problems.

Continuing in this fashion yields the equation

$$P(3, b_3) = \underset{\substack{x_3 = 0, 1 \\ 5x_3 \leqslant b_3}}{\text{maximum}} \{2x_3 + P(4, b_3 - 5x_3)\}$$

where b_3 is given by $b_3 = b_2 - 6x_2$. Also,

$$P(4, b_4) = \underset{\substack{x_4 = 0, 1 \\ 3x_4 \leqslant b_4}}{\text{maximum}} \{7x_4 + P(5, b_4 - 3x_4)\}$$

where $b_4 = b_3 - 5x_3$. Finally,

$$P(5, b_5) = \underset{\substack{x_5 = 0, 1 \\ 7x_5 \leqslant b_5}}{\text{maximum}} \{5x_5\}$$

where $b_5 = b_4 - 3x_4$.

At this point the original problem is converted into a sequence of five problems, $P(1, 15)$, $P(2, b_2)$, $P(3, b_3)$, $P(4, b_4)$, and $P(5, b_5)$. Notice that the last of these problems is the easiest to solve. Very simply, if b_5 is less than 7, then project 5 cannot be funded. If b_5 is 7 or larger, then project 5 can and should be funded. Since it seems so easy to solve the last problem, that of finding $P(5, b_5)$, let us solve the sequence of five problems by working backward. That is, first find $P(5, b_5)$, then $P(4, b_4), \ldots$, until $P(1, 15)$ is found.

There is one minor difficulty with finding $P(5, b_5)$, namely, that the value of b_5 is unknown. For this reason we list all possibilities for b_5, keeping

Table 8.3. Capital Budgeting Problem: Optimal Funding of Project 5

	Reward from state 5 if		
b_5	$x_5 = 0$	$x_5 = 1$	$P(5, b_5)$
0	0	—	0
1	0	—	0
2	0	—	0
3	0	—	0
4	0	—	0
5	0	—	0
6	0	—	0
7	0	5	5
8	0	5	5
9	0	5	5
10	0	5	5
11	0	5	5
12	0	5	5
13	0	5	5
14	0	5	5
15	0	5	5

track of the solutions to the 16 problems given by $P(5, b_5)$, $b_5 = 0, 1, 2, \ldots, 15$. To enable us to keep track of these solutions, they are recorded in Table 8.3. The entries in the table are $v_5 x_5$ for $x_5 = 0$ or 1, and the last column is the larger of the entries. Where no entry appears, project 5 cannot be budgeted.

The next problem to be solved is that of finding $P(4, b_4)$. As before, the value of b_4 is unknown, so 16 different problems are solved, with b_4 assuming values from 0 to 15. For any one of these problems the solution is rather simple. If project 4 is funded, then the value is $7 + P(5, b_4 - 3)$ where $P(5, b_4 - 3)$ can be found in Table 8.3. If project 4 is not funded, then the value is $P(5, b_4)$, which again can be found in Table 8.3. Finally, the larger of $7 + P(5, b_4 - 3)$ and $P(5, b_4)$ is $P(4, b_4)$. For example, if b_4 is 8, then for $x_4 = 0$ the value of $0 + P(5, 8) = 0 + 5 = 5$, while if $x_4 = 1$, the value is $7 + P(5, 8 - 3) = 7 + P(5, 5) = 7 + 0 = 7$. Since 7 is larger than 5, $P(4, 8)$ is equal to 7. These computations as well as the computations for all the different b_4's can be found in Table 8.4.

Next, compute $P(3, b_3)$ in the same way for all values of b_3 between 0 and 15. This time use the identity

$$P(3, b_3) = \underset{\substack{x_3 = 0, 1 \\ 5x_3 \leqslant b_3}}{\text{maximum}} \{2x_3 + P(4, b_3 - 5x_3)\}$$

and find $P(4, b_3 - 5x_3)$ from Table 8.4. The values of $P(3, b_3)$ are presented

Table 8.4. Capital Budgeting Problem: Optimal Funding of Projects 4 and 5

	Reward from state 4 if		
b_4	$x_4 = 0$	$x_4 = 1$	$P(4, b_4)$
0	0	—	0
1	0	—	0
2	0	—	0
3	0	7	7
4	0	7	7
5	0	7	7
6	0	7	7
7	5	7	7
8	5	7	7
9	5	7	7
10	5	12	12
11	5	12	12
12	5	12	12
13	5	12	12
14	5	12	12
15	5	12	12

Table 8.5. Capital Budgeting Problem: Optimal Funding of Projects 3, 4, and 5

| | Reward from state 3 if | | |
b_3	$x_3=0$	$x_3=1$	$P(3,b_3)$
0	0	—	0
1	0	—	0
2	0	—	0
3	7	—	7
4	7	—	7
5	7	2	7
6	7	2	7
7	7	2	7
8	7	9	9
9	7	9	9
10	12	9	12
11	12	9	12
12	12	9	12
13	12	9	12
14	12	9	12
15	12	14	14

in Table 8.5. The 16 values for $P(2,b_2)$ computed from Table 8.5 are shown in Table 8.6.

Finally, the original problem given by Equation (8.1) is solved in order to find $P(1,15)$. Since the budget is known for the five projects, we need to solve only one problem, $P(1,15)$, rather than 16 problems, $P(1,b_1)$ for

Table 8.6. Capital Budgeting Program: Optimal Funding of Projects 2, 3, 4, and 5

| | Reward from state 2 if | | |
b_2	$x_2=0$	$x_2=1$	$P(2,b_2)$
0	0	—	0
1	0	—	0
2	0	—	0
3	7	—	7
4	7	—	7
5	7	—	7
6	7	6	7
7	7	6	7
8	9	6	9
9	9	13	13
10	12	13	13
11	12	13	13
12	12	13	13
13	12	13	13
14	12	15	15
15	14	15	15

budgets b_1 ranging from 0 to 15. From Equation (8.1)

$$P(1,15)=\text{maximum}\{8+P(2,11),0+P(2,15)\}$$
$$=\text{maximum}\{8+13,0+15\} \qquad \text{(using Table 8.6)}$$
$$=\text{maximum}\{21,15\}$$
$$=21$$

Hence, the optimal value for the original problem is 21.

Now that the optimal value is known, the next step is to find which projects yield this value. This time the computations are performed in the forward direction. That is, we begin with $P(1,15)$ and end with $P(5, b_5)$ for some known b_5.

In the set of equations above, the optimal value of 21 is generated partially by $x_1 = 1$. Thus $x_1 = 1$ is part of the optimal solution and $b_2 = 15 - 4 = 11$. In Table 8.6 in the row $b_2 = 11$ the better value is 13 and this value appears in the $x_2 = 1$ column. This means that $x_2 = 1$ is part of the optimal solution and therefore $b_3 = 11 - 6 = 5$. In Table 8.5 for the row $b_3 = 5$ the optimal value is 7 and is given by $x_3 = 0$, which implies that $x_3 = 0$ is part of the optimal solution and $b_4 = 5$. From Table 8.4, $x_4 = 1$ is also part of the optimal solution and yields the value 7, and b_5 is equal to $5 - 3 = 2$. Lastly, in Table 8.3 for b_5 having the value 2, $x_5 = 0$ is the last part of the optimal solution. In summary, the solution is to fund projects 1, 2, and 4 at a cost of $13,000 and with a value equal to 21. This, of course, agrees with the solution found by complete enumeration.

In this case it is optimal not to spend the entire budget. With the information in Tables 8.3 to 8.6 it is very easy to compute the optimal allocations for any budget from 0 through 15. This is done by creating a table for $P(1, b_1)$ with b_1 varying from 0 to 15 and noting that

$$P(1, b_1)= \underset{\substack{x_1=0,1 \\ 4x_1 \leqslant b_1}}{\text{maximum}} \{8x_1 + P(2, b_1 - 4x_1)\}$$

The results are presented in Table 8.7.

Another potential advantage of these tables is that if a sixth proposal is added, the answer can be found immediately. For example, if the new project (numbered 0) has a $5000 cost and a value of 7, then for the original $15,000 budget

$$P(0,15)=\text{maximum}\{0+P(1,15),7+P(1,10)\}$$
$$=\text{maximum}\{0+21,7+15\}$$
$$=\text{maximum}\{21,22\}$$
$$=22$$

Table 8.7. Capital Budgeting Program: Optimal Funding of All Projects

	Reward from state 1 if		
b_1	$x_1 = 0$	$x_1 = 1$	$P(1, b_1)$
0	0	—	0
1	0	—	0
2	0	—	0
3	7	—	7
4	7	8	8
5	7	8	8
6	7	8	8
7	7	15	15
8	9	15	15
9	13	15	15
10	13	15	15
11	13	15	15
12	13	17	17
13	13	21	21
14	15	21	21
15	15	21	21

Working backward through the tables we find that the optimal solution is

$$x_0 = 1 \quad b_1 = 15 - 5 = 10$$
$$x_1 = 1 \quad b_2 = 10 - 4 = 6$$
$$x_2 = 0 \quad b_3 = 6 - 0 = 6$$
$$x_3 = 0 \quad b_4 = 6 - 0 = 6$$
$$x_4 = 1 \quad b_5 = 6 - 3 = 3$$
$$x_5 = 0$$

In summary the dynamic programming approach comprises three steps:

1. Decompose the original five-variable problem into a sequence of five sets of smaller problems such that each has only one variable and is related to the others via the constraints.
2. Solve the sequence of sets of problems beginning with the last problem first.
3. Construct the optimal values of the variables working forward from the first problem to the last.

We generalize this approach in Section 8.2.

8.2. The Principle of Dynamic Programming

To understand the general dynamic programming approach some terminology is needed. Assume that there are n variables $x_1, x_2, \ldots, x_n$ and that the problem is to be solved one variable at a time beginning with x_1. Then there are n subproblems to be solved. Define each of these subproblems as a stage. Thus, the sample problem of Section 8.1 is a five-stage problem. In the sample problem at each stage the concern is with the amount of money that is available. For any stage, this amount could have been 0 through $15,000. The numbers $0, 1, 2, \ldots, 15$ are defined as the *states* associated with the stage. For example, if the funding of projects 1 and 2 requires $10,000 of the $15,000, then it is said that "We are in state 5 at stage 3" since there are $5000 left, and projects 3, 4, and 5 remain for potential funding.

At each stage a decision must be made. Let $D_j(s)$ be the set of possible decisions when in state s at stage j. In the capital budgeting example if the state s represents enough money to fund project j, then the decision is either fund project j or do not fund project j. However, if s is not large enough, then the only possible decision is not to fund project j. When a decision is chosen, a reward accrues and/or a new state is entered at the next stage. The reward may be zero and the new state may be the same as the old state. Let $r_j(d)$ be the reward for making decision d at stage j. Also, let $t_j(s, d)$ be the new state when the old state is s, the stage is j, and the decision is d. For the example it follows that

$$r_j(d) = \begin{cases} 0 & \text{if } d \text{ is "do not fund"} \\ v_j & \text{if } d \text{ is "fund"} \end{cases}$$

and

$$t_j(s, d) = \begin{cases} s & \text{if } d \text{ is "do not fund"} \\ s - c_j & \text{if } d \text{ is "fund"} \end{cases}$$

As we noted earlier, the key to dynamic programming is that the problem is decomposed into smaller problems. This is stated as Bellman's principle of optimality: "An optimal policy has the property that whatever the initial state and decision are, the remaining decisions must constitute an optimal policy with regard to the state resulting from the first decision" (Bellman and Dreyfus, 1962). In other words, if we are in state s at stage j, all remaining decisions must be optimal regardless of how we arrived at state s. Mathematically this is expressed as

$$P(j, s) = \underset{d \in D_j(s)}{\text{maximum}} \{ r_j(d) + P(j+1, t_j(s, d)) \} \qquad (8.2)$$

That is, the solution to the problem at stage j is to make the decision that maximizes the immediate reward r *and* the rewards in future stages $j+1, \ldots, n$ starting in the new state $t_j(s, d)$. Equation (8.2) is termed the *functional equation* or *recursive equation* of a dynamic programming problem.

Finally, although conceptually the problem progresses from stage 1 to stage 2... to stage n, the problem is solved by working backward from stage n to stage $n-1$... to stage 1. This is termed *backward recursion*. As in the sample problem, this procedure gives the optimal value of the objective function but not the optimal values of the variables. These are found by a

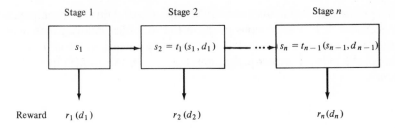

Figure 8.1. Dynamic Programming

forward recursion procedure. Beginning at stage 1, we find the decision that yields the optimal value at each stage.

In Figure 8.1 a pictorial representation of the dynamic programming approach is given. Again, begin in state s_1 and make a decision d_1, thereby bringing about a new state $s_2 = t_1(s_1, d_1)$ and a reward $r_1(d_1)$. Then make the decision d_2, which results in a new state $s_3 = t_2(s_2, d_2)$, and again collect a reward $r_2(d_2)$. This continues until the terminal stage n.

8.3. Dynamic Programming Problems

In this section several standard types of problems that can be solved by the dynamic programming method are presented. Although each type can be solved by using dynamic programming, the approach is generally different for each type of problem. The key points are to identify the states, stages, and transition function in order to write the functional equation.

8.3.1. Resource Allocation Problems

Consider the following problem: A camper is loading his knapsack with cans of food for his trip. He can carry only 20 pounds of food. The following table shows the weights of the four different types of food the camper is considering and the value he places on each type of food.

Food i	Weight w_i	value v_i
1	7	12
2	3	5
3	8	10
4	4	6

Notice that the problem can be expressed as

$$\text{Maximize} \quad 12x_1 + 5x_2 + 10x_3 + 6x_4$$
$$\text{subject to} \quad 7x_1 + 3x_2 + 8x_3 + 4x_4 \leqslant 20$$
$$x_j \geqslant 0 \quad \text{and} \quad x_j \text{ integer} \quad j = 1, 2, 3, 4$$

This problem is a generalization of the capital budgeting problem presented in Section 8.1. In the previous problem the variables could assume only the value of 0 or 1, whereas in this example any integer value is permissible. This problem is an integer programming problem. A problem in the general form

$$\text{Maximize} \quad \sum_{j=1}^{n} v_j x_j$$

$$\text{subject to} \quad \sum_{j=1}^{n} w_{ij} x_j \leqslant b_i \qquad i=1,2,\ldots,m$$

$$x_j \geqslant 0 \quad \text{and} \quad x_j \text{ integer} \qquad j=1,2,\ldots,n$$

is termed an m-dimensional knapsack problem due to the interpretation given above. That is, a linear programming resource allocation problem is called a knapsack problem when the variables must be integers. Notice that a capital budgeting problem is a special case of the knapsack problem. The knapsack example is a one-dimensional problem, since there is only one constraint. The procedure for solving this problem is similar to the previous approach in the sense that we solve for one variable at a time. Let $P(j,c)$ represent the optimal value of a knapsack problem with foods $j, j+1,\ldots,4$ and capacity c. The original problem is given by

$$P(1,20) = \underset{\{x_1:\, w_1 x_1 \leqslant 20,\, x_1 \text{ integer}\}}{\text{maximum}} \left\{ v_1 x_1 + P(2, 20 - w_1 x_1) \right\}$$

Furthermore, let $c_2 = 20 - w_1 x_1$; then the second problem is

$$P(2,c_2) = \underset{\{x_2:\, w_2 x_2 \leqslant c_2,\, x_2 \text{ integer}\}}{\text{maximum}} \left\{ v_2 x_2 + P(3, c_2 - w_2 x_2) \right\}$$

and for $c_3 = c_2 - w_2 x_2$

$$P(3,c_3) = \underset{\{x_3:\, w_3 x_3 \leqslant c_3,\, x_3 \text{ integer}\}}{\text{maximum}} \left\{ v_3 x_3 + P(4, c_3 - w_3 x_3) \right\}$$

and for $c_4 = c_3 - w_3 x_3$ find

$$P(4,c_4) = \underset{\{x_4:\, w_4 x_4 \leqslant c_4,\, x_4 \text{ integer}\}}{\text{maximum}} \left\{ v_4 x_4 \right\}$$

Notice that given c_4, the last problem is very easy to solve. The value of c_4 is between 0 and 20 but unfortunately its exact value is unknown. Thus, let us solve the last problem for *all* values of c_4 between 0 and 20. The 21 solutions are presented in Table 8.8.

Now find $P(3,c_3)$ for all possible values of c_3. To create a table for $P(3,c_3)$ simply follow the equation

$$P(3,c_3) = \underset{\{x_3:\, w_3 x_3 \leqslant c_3,\, x_3 \text{ integer}\}}{\text{maximum}} \left\{ v_3 x_3 + P(4, c_3 - w_3 x_3) \right\}$$

Table 8.8. Knapsack Problem: Optimal Loading of Food 4

c_4	x_4	$v_4 x_4$
0	0	0
1	0	0
2	0	0
3	0	0
4	1	6
5	1	6
6	1	6
7	1	6
8	2	12
9	2	12
10	2	12
11	2	12
12	3	18
13	3	18
14	3	18
15	3	18
16	4	24
17	4	24
18	4	24
19	4	24
20	5	30

Thus, for example, if $c_3 = 15$, then x_3 is 0 or 1. If x_3 is 0, then c_4 is 15. If x_3 is 1, then c_4 is $15-8=7$. Hence, the values for $c_3 = 15$ are

$$(x_3=0) \qquad 0 + P(4,15) = 0 + 18 = 18$$

or

$$(x_3=1) \qquad 10 + P(4,7) = 10 + 6 = 16$$

where $P(4, c_4)$ comes from Table 8.8. These computations are summarized in Table 8.9, which presents the answer to all two-variable problems with capacities of anywhere from 0 to 20. Thus, Table 8.9 represents the solution to 21 different problems!

Now solve problem $P(2, c_2)$ for 21 different values of c_2 using the information in Table 8.9. The procedure is very similar to the previous one: for example, if $c_2 = 14$, then x_2 can be 0, 1, 2, 3, or 4 and is presented as

x_2	$v_2 x_2$	$c_2 - x_2 w_2$	$P(3, c_3)$	$v_2 x_2 + P(3, c_3)$
0	0	14	18	18
1	5	11	12	17
2	10	8	12	22[a]
3	15	5	6	21
4	20	2	0	20

[a] This is the largest (optimal) value.

Table 8.9. Knapsack Problem: Optimal Loading of Foods 3 and 4

c_3	$x_3=0$	$x_3=1$	$x_3=2$	$P(3,c_3)$
0	0	—	—	0
1	0	—	—	0
2	0	—	—	0
3	0	—	—	0
4	6	—	—	6
5	6	—	—	6
6	6	—	—	6
7	6	—	—	6
8	12	10	—	12
9	12	10	—	12
10	12	10	—	12
11	12	10	—	12
12	18	16	—	18
13	18	16	—	18
14	18	16	—	18
15	18	16	—	18
16	24	22	20	24
17	24	22	20	24
18	24	22	20	24
19	24	22	20	24
20	30	28	26	30

Since 22 is the largest value, this number is placed in the column $P(2,c_2)$. Again Table 8.10 is constructed by solving 21 problems.

Finally solve $P(1,20)$; this time the weight capacity of the knapsack is known and the results are presented below.

$$P(1,20)=\text{maximum}\{v_1x_1+P(2,20-7x_1)\}$$
$$=\text{maximum}\{0+P(2,20),12+P(2,13),24+P(2,6)\}$$
$$=\text{maximum}\{32,33,34\}=34$$

Thus the optimal solution has a value of 34. In order to find the optimal values for x_1,x_2,x_3,x_4 we have to work backward through the tables. In the last table, since 34 is optimal, $x_1=2$ and $c_2=20-w_1x_1=20-14=6$. Thus in Table 8.10, $v_2(6)=10$ and $x_2=2$. Thus, $c_3=6-6=0$ and $x_3=x_4=0$. The final solution is

$$x_1=2 \quad x_2=2 \quad x_3=0 \quad x_4=0$$

Although the last example is linear in the constraints and objective function, one of the major advantages of dynamic programming is that nonlinear programming problems can also be solved using the same approach without increasing the degree of difficulty. For this reason another examples is considered.

Table 8.10. Knapsack Problem: Optimal Loading of Foods 2, 3, and 4

c_2	0	1	2	3	4	5	6	$P(2, c_2)$
				x_2				
0	0							0
1	0							0
2	0							0
3	0	5						5
4	6	5						6
5	6	5						6
6	6	5	10					10
7	6	11	10					11
8	12	11	10					12
9	12	11	10	15				15
10	12	11	16	15				16
11	12	17	16	15				17
12	18	17	16	15	20			20
13	18	17	16	21	20			21
14	18	17	22	21	20			22
15	18	23	22	21	20	25		25
16	24	23	22	21	26	25		26
17	24	23	22	27	26	25		27
18	24	23	28	27	26	25	30	30
19	24	29	28	27	26	31	30	31
20	30	29	28	27	32	31	30	32

Nonlinear knapsack problem. A company is remodeling its three factories. It has five repairmen who can be sent to the three factories. Each repairman can be assigned to only one factory. The accompanying table shows the values of the repairs when the number of repairmen sent is 0 to 5. Notice that the values of the improvements are not linear in the number of repairmen. That is, on certain operations two men do a job more than twice as well as one man working alone.

	Factory		
	1	2	3
	Atlanta	Boston	Chicago
Number of repairmen	$v_1(x_1)$	$v_2(x_2)$	$v_3(x_3)$
0	0	0	0
1	10	17	15
2	21	35	31
3	38	37	42
4	43	38	49
5	46	38	52

The problem is to determine the number of repairmen who should be sent to each plant in order to maximize the total value. In this problem the stages are the plants and the states are the number of repairmen available, so we decompose this problem according to the plants. Let $P(j, m)$ be the best allocation of repairmen to plants $\{j, j+1, \ldots, 3\}$ when m repairmen are available. Let x_j be the number of repairmen allocated to plant j. It follows that the recursive equation exists.

$$P(1,5) = \underset{x_1=0,1,\ldots,5}{\text{maximum}} \{v_1(x_1) + P(2, 5-x_1)\}$$

Letting $m_2 = 5 - x_1$, we have

$$P(2, m_2) = \underset{x_2=0,1,\ldots,m_2}{\text{maximum}} \{v_2(x_2) + P(3, m_2 - x_2)\}$$

and for $m_3 = m_2 - x_2$ we want to find

$$P(3, m_3) = \underset{x_3=0,1,\ldots,m_3}{\text{maximum}} \{v_3(x_3)\} = v_3(m_3)$$

because $v_3(m_3)$ increases with m_3.

As before, consider all of the possibilities for m_3 and solve for the six different values of m_3. This information is given in Table 8.11.

Now solve $P(2, m_2)$ for all values of m_2 using Table 8.11. Again six problems are solved, but each is relatively easy. For example, if $m_2 = 3$, then

$$P(2,3) = \underset{x_2=0,1,2,3}{\text{maximize}} \{v_2(x_2) + P(3, 3-x_2)\}$$

and for

$$
\begin{aligned}
x_2 = 0 \qquad & v_2(0) + P(3,3) = 0 + 42 = 42 \\
x_2 = 1 \qquad & v_2(1) + P(3,2) = 17 + 31 = 48 \\
x_2 = 2 \qquad & v_2(2) + P(3,1) = 35 + 15 = 50 \\
x_2 = 3 \qquad & v_2(3) + P(3,0) = 37 + 0 = 37
\end{aligned}
$$

$$P(2,3) = \text{maximum}\{42, 48, 50, 37\}$$

Table 8.11. Repairmen Problem: Optimal Allocation to City 3

	Chicago	
m_3 number of servers	x_3	$P(3, m_3) = v_3(m_3)$
0	0	0
1	1	15
2	2	31
3	3	42
4	4	49
5	5	52

Since 50 is the largest value, this is $P(2,3)$. Table 8.12 contains the solution to all six problems.

One final stage must be completed for this problem, and that is to compute $P(1,5)$. Using Table 8.12 and the information in the problem it follows that

Atlanta	x_1					
	0	1	2	3	4	5
5	0+77	10+66	21+50	38+35	43+17	46+0
$v_1(x_1)+P(2,5-x_1)$	77	76	71	73	60	46

Thus the optimal value is 77. Again reverse the direction through the tables in order to find the optimal solution. Since $x_1=0$, then $m_2=5$ and from Table 8.12, $P(2,5)=77$ and $x_2=2$. Thus $m_3=3$ and from Table 8.11, $x_3=3$.

The multidimensional knapsack problem. Consider the knapsack problem with the additional constraint that the knapsack holds only 12 cubic feet of cans. The complete set of data for the problem is given below:

Item	Weight	Volume	Value
1	7	2	12
2	3	3	5
3	8	5	10
4	4	3	6

The problem is now given by

$$\text{Maximize} \quad 12x_1+5x_2+10x_3+6x_4$$
$$\text{subject to} \quad 7x_1+3x_2+8x_3+4x_4\leq 20$$
$$2x_1+3x_2+5x_3+3x_4\leq 12$$
$$x_1,x_2,x_3,x_4\geq 0$$
$$x_1,x_2,x_3,x_4 \text{ integer}$$

Table 8.12. Repairmen Problem: Optimal Allocation to Cities 2 and 3

m_2	Boston: x_2						$P(2,m_2)$
	0	1	2	3	4	5	
0	0						0
1	15	17					17
2	31	32	35				35
3	42	48	50	37			50
4	49	59	66	52	38		66
5	52	66	77	68	53	38	77

Thus, in this two-dimensional knapsack problem two variables are required to describe the state of the system. One variable represents the amount of weight left and the other represents the amount of space left. All possible states that the system can be in are given by $\{(y, z): 0 \leqslant y \leqslant 20, 0 \leqslant z \leqslant 12, y, z \text{ integer}\}$.

One approach to this problem is the conventional dynamic programming method. That is, start by creating a table that has $21 \cdot 13 = 273$ entries where each entry is $P(4, (b_1, b_2))$—the optimal solution to the knapsack problem with only item 4 and an amount of weight given by b_1 and volume given by b_2. Such a table would look like Table 8.13. Even in this two-dimensional problem the first problem is trivial to solve. Unfortunately, the bookkeeping for the remaining problems becomes very tedious. To create a table for the problem $P(3, (b_1, b_2))$, for each of the 273 entries:

1. Find m, the largest amount of food 3 that can be used if the resources are available in amounts b_1 and b_2;
2. For all integers from 0 to m, let $x_3 = 0, 1, \ldots, m$ and compute the value
   $$10x_3 + P(4, (b_1 - 8x_3, b_2 - 5x_3));$$
3. Find the maximum of all of these values.

Table 8.13. Multidimensional Knapsack Problem: Optimal Loading of Food 4

Weight	0 1 2	3 4 5	6 7 8	9 10 11	12
0					
1	—	—	—	—	—
2					
3					
4					
5	—	6	6	6	6
6					
7					
8					
9					
10	—	6	12	12	12
11					
12					
13					
14	—	6	12	12	12
15					
16					
17					
18	—	6	12	12	12
19					
20	—	6	12	12	12

Table 8.14. Multidimensional Knapsack Problem: One-Item Feasible Loadings

S_1	S_2	S_3	S_4
(0;0,0)	(0;0,0)	(0;0,0)	(0;0,0)
(12;7,2)	(5;3,3)	(10;8,5)	(6;4,3)
(24;14,4)	(10;6,6)	(20;16,10)	(12;8,6)
	(15;9,9)		(18,12,9)
	(20;12,12)		(24;16,12)

All of this work is easy to perform but there is so much of it that it makes dynamic programming very unappealing. Furthermore, if the problem has three constraints it is difficult to write out the information in a three-dimensional table. Therefore, we suggest the following approach, which is dynamic programming but with an improved bookkeeping process. In fact, the approach presented works for any knapsack problem with nonnegative coefficients in the constraints and objective function.

The technique is presented step by step, using the two-dimensional knapsack problem as an example.

1. Let the triplet $P_j(a; b, c)$ denote a solution with value a, using b units of resource 1 and c units of resource 2 with only products $j, j+1, \ldots, n$. For the example, then, $P_2(7; 12, 5)$ would indicate that items 2, 3, and 4 can be loaded such that the weight used is 12, the volume used is 5, and the value is 7.

2. Let P_j denote the set of all *feasible* loadings of items $j, j+1, \ldots, n$.

3. Let S_j denote the set of all feasible loadings using only item j. For the example, note that up to 2 cans of food 1, up to 4 cans of food 2, up to 2 cans of food 3, and up to 4 cans of food 4 can be loaded. The elements of the sets S_j are presented in Table 8.14.

The goal is to find P_1, the set of all feasible loadings of products 1, 2, 3, and 4, and then to select the loading in P_1 that has the greatest value. This is done by finding the recursive relationship that will bring us from P_4 to P_3 to P_2 and finally to P_1.

Table 8.15. Multidimensional Knapsack Problem: Loadings of Items 3 and 4

	S_3		
	0 cans	1 can	2 cans
P_4	(0;0,0)	(10;8,5)	(20;16,10)
(0;0,0)	(0;0,0)	(10;8,5)	(20;16,10)
(6;4,3)	(6;4,3)	(16;12,8)	(26;20,13)[a]
(12;8,6)	(12;8,6)	(22;16,11)	(32;24,16)[a]
(18;12,9)	(18;12,9)	(28;20,14)[a]	(38;28,19)[a]
(24;16,12)	(24;16,12)	(34;24,17)[a]	(44;32,22)[a]

[a] This solution is infeasible.

Notice that P_4 is identical to S_4. In order to find P_3, use a two-step process. The first step is to add each element of S_3 to each element of P_4. This is done in Table 8.15. The addition is done element by element. For example, add (10; 8,5) and (6; 4,3) to obtain (16; 12,8). The second step is to eliminate from this table those entries that are not feasible. The results can be seen in Table 8.15. The infeasible entries are identified in the footnote, and the remaining feasible entries form the set P_3. This step is generally written as

$$\text{step 4} \qquad P_{j-1} = P_j \oplus S_{j-1}$$

where it is understood that the $\oplus$ operation means, "Add every element in P_j to every element in S_{j-1} *and* check for feasibility." Hence the remainder of the method is to perform step 4 for $j = n, n-1, \dots, 2$. Finally, after computing P_1, find the solution that yields the largest value.

Continuing with the example, we note that the feasible loadings from Table 8.15 are listed as P_3 in Table 8.16. The columns in Table 8.16 are the elements of S_2, and performing the addition and feasibility test yields P_2. Notice that if we add by rows there is no need to perform all of the additions; for example, the result of adding (12; 8,6) and (15; 9,9) is (27; 17,15). Since this solution is not feasible, it follows that (12; 8,6) and (20; 12,12) is also not feasible.

The last table to be computed is Table 8.17, where P_2 is taken from Table 8.16 and added to S_1. Observe that the largest value of a feasible solution is 34, given by (24; 14,4) from S_1 and (10; 6,6) from P_2. Now work backward through the tables, beginning with two cans of food 1 from Table 8.17, to find the actual loadings. We see that the best element of P_2 is (10; 6,6) and referring back to Table 8.16 find that (10; 6,6) appears in the (0; 0,0) row and (10; 6,6) column; hence load 2 cans of food 2. Since P_3 is (0; 0,0), refer to Table 8.15 and find that (0; 0,0) appears in the (0; 0,0) row and (0; 0,0) column, so do not load foods 3 or 4.

Table 8.16. Multidimensional Knapsack Problem: Loadings of Items 2, 3, and 4

	S_2				
P_3	0 cans (0;0,0)	1 can (5;3,3)	2 cans (10;6,6)	3 cans (15;9,9)	4 cans (20;12,12)
(0;0,0)	(0;0,0)	(5;3,3)	(10;6,6)	(15;9,9)	(20;12,12)
(6;4,3)	(6;4,3)	(11;7,6)	(16;10,9)	(21;13,12)	(26;16,15)[a]
(12;8,6)	(12;8,6)	(17;11,9)	(22;14,12)	(27;17,15)[a]	
(18;12,9)	(18;12,9)	(23;15,12)	(28;18,15)[a]		
(24;16,12)	(24;16,12)	(29;19,15)[a]			
(10;8,5)	(10;8,5)	(15;11,8)	(20;14,11)	(25;17,14)[a]	
(16;12,8)	(16;12,8)	(21;15,11)	(26;18,14)[a]		
(22;16,11)	(22;16,11)	(27;19,14)[a]			
(20;16,10)	(20;16,10)	(25;19,13)[a]			

[a] This solution is infeasible.

Table 8.17. Multidimensional Knapsack Problem: Four-Item Loadings

P_2	0 cans (0;0,0)	1 can (12;7,2)	2 cans (24;14,4)
(0;0,0)	(0;0,0)	(12;7,2)	(24;14,4)
(6;4,3)	(6;4,3)	(18;11,5)	(30;18,7)
(12;8,6)	(12;8,6)	(24;15,8)	(36;22,10)^a
(18;12,9)	(18;12,9)	(30;19,11)	(42;26,13)^a
(24;16,12)	(24;16,12)	(36;23,14)^a	
(10;8,5)	(10;8,5)	(22;15,7)	(34;22,9)^a
(16;12,8)	(16;12,8)	(28;19,10)	(40;26,12)^a
(22;16,11)	(22;16,11)	(34;23,13)^a	
(20;16,10)	(20;16,10)	(32;23,12)^a	
(5;3,3)	(5;3,3)	(17;10,5)	(29;17,7)
(11;7,6)	(11;7,6)	(23;14,8)	(35;21,10)^a
(17;11,9)	(17;11,9)	(29;18,11)	(41;25,13)^a
(23;15,12)	(23;15,12)	(35;22,14)^a	
(15;11,8)	(15;11,8)	(27;18,10)	(39;25,12)^a
(21;15,11)	(21;15,11)	(33;22,13)^a	
(10;6,6)	(10;6,6)	(22;13,8)	(34;20,10)^b
(16;10,9)	(16;10,9)	(28;17,11)	(40;24,13)^a
(22;14,12)	(22;14,12)	(34;21,14)^a	
(20;14,11)	(20;14,11)	(32;21,13)^a	
(15;9,9)	(15;9,9)	(27;16,11)	(39;23,13)^a
(21;13,12)	(21;13,12)	(33;20,14)^a	
(20;12,12)	(20;12,12)	(32;19,14)^a	

^{a}This solution is infeasible.
^{b}This is the optimal solution.

8.3.2. Production Planning Problems

Consider the following production planning problem. The Nurit Sneaker Company has forecast demands for sneakers for the next 4 months. These forecasts are given, in thousands of pairs, in Table 8.18. The production manager must have a production schedule immediately. That is, he must know how many pairs of sneakers are to be produced during each month from January through April. A schedule then is represented by (x_1, x_2, x_3, x_4) where x_j represents the number of pairs, in thousands, produced during period j. The schedule is feasible if the demands are always met. That is,

$$x_1 \geqslant 4$$
$$x_1 + x_2 \geqslant 5$$
$$x_1 + x_2 + x_3 \geqslant 8$$
$$x_1 + x_2 + x_3 + x_4 = 10$$

The last constraint is expressed as an equality since there is no reason to produce more sneakers than are demanded. (Assume that this product is

Table 8.18

	Month	Demand[a]
January	1	4
February	2	1
March	3	3
April	4	2

[a] In thousands of pairs.

Table 8.19. Production Planning by Explicit Enumeration[a]

Production level in				Production level in				Production level in			
A	B	C	D	A	B	C	D	A	B	C	D
4	1	3	2	5	0	3	2	6	0	2	2
4	1	4	1	5	0	4	1	6	0	3	1
4	1	5	0	5	0	5	0	6	0	4	0
4	2	2	2	5	1	2	2	6	1	1	2
4	2	3	1	5	1	3	1	6	1	2	1
4	2	4	0	5	1	4	0	6	1	3	0
4	3	1	2	5	2	1	2	6	2	0	2
4	3	2	1	5	2	2	1	6	2	1	1
4	3	3	0	5	2	3	0	6	2	2	0
4	4	0	2	5	3	0	2	6	3	0	1
4	4	1	1	5	3	1	1	6	3	1	0
4	4	2	0	5	3	2	0	6	4	0	0
4	5	0	1	5	4	0	1				
4	5	1	0	5	4	1	0				
4	6	0	0	5	5	0	0				

Production level in				Production level in			
A	B	C	D	A	B	C	D
7	0	1	2	8	0	0	2
7	0	2	1	8	0	1	1
7	0	3	0	8	0	2	0
7	1	0	2	8	1	0	1
7	1	1	1	8	1	1	0
7	1	2	0	8	2	0	0
7	2	0	1	9	0	0	1
7	2	1	0	9	0	1	0
7	3	0	0	9	1	0	0
				10	9	9	0

[a] A, B, C, and D stand for January, February, March, and April, respectively; numbers in the table body represent thousands of pairs of sneakers.

replaced in May by an improved sneaker.) The number of production schedules that satisfy the constraints is 61. These are listed in Table 8.19 simply to demonstrate that even for this small problem there are a fairly large number of possible answers, and the task of identifying these answers is not easy. There are costs associated with any production schedule. The costs are $7 per pair produced, $40,000 for each month that has a production run (setup cost), and $10 per month for each pair that is produced but shipped later. For example, the total cost in thousands of dollars due to the schedule $(5, 0, 3, 2)$ would be $5 \cdot 7 + 0 \cdot 7 + 3 \cdot 7 + 2 \cdot 7$ for the units produced plus $3 \cdot 40$ for the three setups in January, March, and April, plus $1 \cdot 10$ for the sneakers produced in January and sold in February.

This problem can be formulated as a dynamic programming problem. Let z_i be the number of units on hand at the beginning of month (stage) i for $i = 1, 2, 3, 4$. Then

$$z_1 = 0 \qquad z_2 = x_1 - 4 \qquad z_3 = z_2 + x_2 - 1 \qquad z_4 = z_3 + x_3 - 3$$

and

$$z_5 = z_4 + x_4 - 2$$

which must be zero. The functional equation then becomes

$$P_j(z_j) = \min_{\substack{\{x_j : z_j + x_j \geq d_j; \\ z_j + x_j \leq d_j + d_{j+1} + \cdots + d_n\}}} \{7x_j + 10(z_j + x_j - d_j) + 40\delta(x_j) + P_{j+1}(z_j + x_j - d_j)\}$$

$$\quad \text{(satisfy demand)}$$
$$\quad \text{(do not overproduce)}$$

where

$$\delta(x_j) = \begin{cases} 1 & \text{if } x_j > 0 \\ 0 & \text{otherwise} \end{cases}$$

Working backward, we find that z_4 can be either 0, 1, or 2, in which case x_4 must equal $2 - z_4$. The costs are given in Table 8.20. For the problem of finding $P_3(z_3)$ note that z_3 can be any amount from 0 to 5. Notice that for a given z_3, the amount produced, x_3, is bounded above and below by the two facts that

$$x_3 + z_3 \leq 5 \quad \text{or} \quad x_3 \leq 5 - z_3$$

Table 8.20. Production Planning: Optimal Production for April

z_4	x_4			$P_4(z_4)$
	0	1	2	
0	×	×	54	54
1	×	47	×	47
2	0	×	×	0

× indicates an infeasible solution

Table 8.21. Production Planning: Optimal Production for March and April

				x_3				
z_3	0	1	2	3	4	5	$P_3(z_3)$
0	×	×	×	115	125	95	95
1	×	×	108	118	88	×	88
2	×	101	111	81	×	×	81
3	54	104	74	×	×	×	54
4	57	67	×	×	×	×	57
5	20	×	×	×	×	×	20

× indicates an infeasible solution

and

$$x_3 + z_3 \geqslant 3 \quad \text{or} \quad x_3 \geqslant 3 - z_3$$

Now compute the cost table for $P_3(z_3)$. These costs are shown in Table 8.21; they are given by the equation

$$P_3(z_3) = \begin{cases} 10(z_3 - 3) + P_4(z_3 - 3) & \text{if } x_3 = 0 \\ 40 + 7x_3 + 10(z_3 + x_3 - 3) + P_4(z_3 + x_3 - 3) & \text{if } x_3 > 0 \end{cases}$$

For example, if z_3 equals 2, then x_3 is 1, 2, or 3. For x_3 equal to 1 the total costs are 40 (setup) + 7 (unit cost) + 0 (holding) + 54 ($P_4(0)$), which totals 101. For $x_3 = 2$ the total costs are 40 (setup) + 14 (unit cost) + $10 (holding one item) + 47 ($P_4(1)$), which totals 111. Similarly, for x_3 equal to 3 the total cost is 81 and since 81 is the smallest value, this is $P_3(2)$.

Using Table 8.21, compute $P_2(z_2)$ and notice that z_2 and x_2 can each vary from 0 to 6 subject to the constraints that $z_2 + x_2 \geqslant 1$ and $z_2 + x_2 \leqslant 6$. The computations for $P_2(z_2)$ can be found in Table 8.22. Finally, compute $P_1(z_1)$; however, z_1 is zero, so we do not need an entire table, but simply one row. Also, x_1 is at least 4, so x_1 varies from 4 to 10 and the cost when we

Table 8.22. Production Planning: Optimal Production for February, March, and April

				x_2				
z_2	0	1	2	3	4	5	6	$P_2(z_2)$
0	×	142	152	162	152	172	152	142
1	95	145	158	148	165	145	×	95
2	98	148	138	158	138	×	×	98
3	101	131	151	114	×	×	×	101
4	84	144	134	×	×	×	×	84
5	97	117	×	×	×	×	×	97
6	70	×	×	×	×	×	x	70

× indicates an infeasible solution

choose the value x_1 is given by

$$c(x_1)=40+7x_1+10(x_1-4)+P_2(x_1-4)$$

Using this equation and Table 8.22 we find that

x_1	4	5	6	7	8	9	10	$P_1(0)$
cost	210	180	210	220	220	250	240	180

From this row notice that 180 is the minimum cost that $x_1=5$, which results in $z_2=1$. From Table 8.22, $P_2(z_2)=95$, so that $x_2=0$ and $z_3=2-2=0$. From Table 8.21, $P_3(0)=95$ and $x_3=5$, which yields $z_4=5-5=0$ and $x_4=0$. The optimal schedule is $(5,0,5,0)$ at a minimum cost of \$180.

The general production planning problem. One useful feature of dynamic programming is that typically if a problem can be solved by means of dynamic programming, then a more general problem can also be solved without much change. For example, let n be the number of periods in a production planning problem. Suppose that the costs vary from month to month. That is, k_i is the setup cost if there is production during month i, h_i is the per unit holding cost for units left over at the end of the month i, and c_i is the per unit production cost for items made during month i. Then using the dynamic programming approach we have

$$P_j(z_j)=\text{minimum}\{k_j\delta(x_j)+c_jx_j+h_j(z_j+x_j-d_j)+P_{j+1}(z_j+x_j-d_j)\}$$

where again

$$\delta(x_j)=\begin{cases}1 & \text{if } x_j>0 \\ 0 & \text{if } x_j=0\end{cases}$$

8.3.3. Shortest Path Problems

In Figure 8.2 a graphical representation of a traffic network is presented. There are seven locations and assorted roads between pairs of locations. The numbers along the roads represent the time required (in hours) to travel the roads. Notice that timewise the shortest path between two points is not necessarily a straight line. For example, the shortest path from C to F is C to E to F, not C to F. One problem is to find the path between A and G that takes the least amount of time to travel. In fact, using dynamic programming it is possible to find the shortest path from each city to every other city.

Table 8.23 presents the travel times in tabular form. In general the values can be time, distance, cost, or other measures. The entry in row i, column j,

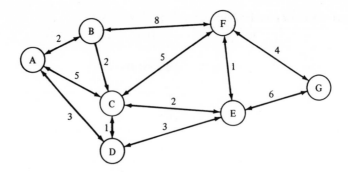

Figure 8.2. Shortest Route Problem

is the time required to travel along the road ij. If no road exists, a time of M is used, where M is a very large number, and for roads i to i the length of time is zero.

Now consider the following problem: What is the minimum amount of time required to travel from any city i to any city j using at most two roads? If two roads are used, then the possible paths from i to j are all represented by i to k to j where k is either A, B, C, D, E, F, or G. The minimum then is the minimum of $\{d_{ik}+d_{kj}, k=A,B,\ldots,G\}$ where d_{ik} and d_{kj} both come from Table 8.23. For example, the minimum two-step (or fewer) path from C to G is found by evaluating:

Path	Time
C to A to G	$5+M=M$
C to B to G	$2+M=M$
C to C to G[a]	$0+M=M$
C to D to G	$1+M=M$
C to E to G	$2+6=8$
C to F to G	$5+4=9$
C to G to G[a]	$M+0=M$

[a] This is a one-step path.

Table 8.23. Shortest Path Problem: One-Step Distance Matrix

	A	B	C	D	E	F	G
A	0	2	5	3	—	—	—
B	2	0	2	—	—	8	—
C	5	2	0	1	2	5	—
D	3	—	1	0	3	—	—
E	—	—	2	3	0	1	6
F	—	8	5	—	1	0	4
G	—	—	—	—	6	4	0

Table 8.24. Shortest Path Problem: Two-Step Distance Matrix

	A	B	C	D	E	F	G
A	0	2	4	3	6	10	—
B	2	0	2	3	4	7	12
C	4	2	0	1	2	3	8
D	3	3	1	0	3	4	9
E	6	4	2	3	0	1	5
F	10	7	3	4	1	0	4
G	—	12	8	9	5	4	0

and noting that the minimum is 8. This computation is performed for every pair of cities, and the minimum times can be found in Table 8.24. Call the table $D(2)$.

Next compute the minimum times when using three roads or fewer. Again decompose the trip into a two-step process. That is, for any pair i to j to travel on three roads means to go from i to k on one road and from k to j on two roads. Thus if $D_{ij}(3)$ represents the least amount of time to go from i to j using at most three roads, then

$$D_{ij}(3) = \underset{k}{\text{minimum}} \left\{ d_{ik} + D_{kj}(2) \right\}$$

where d_{ik} comes from the original table of times (Table 8.23) and $D_{kj}(2)$ comes from the table of two-road times. For example, for cities B and F it follows that

	Path	
One road	Two roads	
B to A	A to F	$2 + 10 = 12$
B to B	B to F	$0 + 7 = 7$
B to C	C to F	$2 + 3 = 5$
B to D	D to F	$M + 4 = M$
B to E	E to F	$M + 1 = M$
B to F	F to F	$8 + 0 = 8$
B to G	G to F	$M + 4 = M$

Table 8.25. Shortest Path Problem: Three-Step Distance Matrix

	A	B	C	D	E	F	G
A	0	2	4	3	6	7	12
B	2	0	2	3	4	5	10
C	4	2	0	1	2	3	7
D	3	3	1	0	3	4	8
E	6	4	2	3	0	1	5
F	7	5	3	4	1	0	4
G	12	10	7	8	5	4	0

Table 8.26. Shortest Path Problem: Four-Step Distance Matrix

	A	B	C	D	E	F	G
A	0	2	4	3	6	7	11
B	2	0	2	3	4	5	9
C	4	2	0	1	2	3	7
D	3	3	1	0	3	4	8
E	6	4	2	3	0	1	5
F	7	5	3	5	1	0	4
G	11	9	7	8	5	4	0

Table 8.27. Shortest Path Problem: Distance Matrix for Five or More Steps

	A	B	C	D	E	F	G
A	0	2	4	3	6	7	11
B	2	0	2	3	4	5	9
C	4	2	0	1	2	3	7
D	3	3	1	0	3	4	8
E	6	4	2	3	0	1	5
F	7	5	3	4	1	0	4
G	11	9	7	8	5	4	0

The minimum is 5 and this can be found in Table 8.25, along with all of the three-road minimums. In general, the functional equation is given by

$$D_{ij}(r+1) = \operatorname*{minimum}_{k}\{d_{ik} + D_{kj}(r)\} \qquad r = 1, 2, \ldots, n-1$$

where r is the number of roads permitted. Furthermore, no optimal path visits each city more than once. Thus, all that is needed is to repeat the process until $D_{ij}(7)$ is computed and the minimum times required to travel from city i to j appear in $D_{ij}(7)$. We compute $D_{ij}(4)$ and $D_{ij}(5)$; these values are given in Tables 8.26 and 8.27, respectively. Also, $D_{ij}(6)$ is identical to $D_{ij}(5)$, which means that no improvements can be made. Hence, Table 8.27 presents the distance of the shortest path between any two cities.

8.4. Summary

In this chapter problems are examined that are more general in structure than either the integer or linear programming problems examined in previous chapters. In fact, there is no unique form of the dynamic programming problem, as can be seen from the varied examples in the chapter.

All of the problems do have the property that they can be decomposed into *stages* and solved one stage at a time. That is, the problems all satisfy *Bellman's principle of optimality*. Furthermore, the solution is obtained conceptually by working forward but computationally by the *backward recursion procedure*.

Some types of problems for which dynamic programming has been extensively used are *capital budgeting, knapsack, shortest distance,* and *production planning.* There are many other applications of dynamic programming; they are discussed in the books in the reference list for this chapter.

References and Selected Readings

Bellman, R. 1957. *Dynamic Programming.* Princeton, New Jersey: Princeton University Press.

Bellman, R., and S. E. Dreyfus 1962. *Applied Dynamic Programming.* Princeton, New Jersey: Princeton University Press.

Hadley, G. 1964. *Nonlinear and Dynamic Programming.* Reading, Massachusetts, Addison-Wesley.

Howard, R. A. 1960. *Dynamic Programming and Markov Processes.* New York: John Wiley and Sons, Inc.

Nemhauser, G. 1966. *Introduction to Dynamic Programming.* New York: John Wiley and Sons, Inc.

Problems

1. In the two networks given here find the shortest route from city A to city B by using dynamic programming.

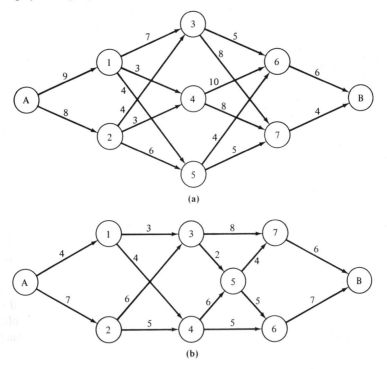

(a)

(b)

2. Pennsylvania Pottery wants to allocate 10 daily advertisements to 4 different radio stations. The effectiveness, measured in number of people reached, of the number of daily ads assigned to each station is shown in the accompanying table. Find the optimal allocation.

Radio stations	Number of advertisements									
	1	2	3	4	5	6	7	8	9	10
WKRB	6	10	12	20	22	23	23	24	24	25
WABD	4	8	10	15	18	20	21	21	23	23
KRUP	5	9	15	20	20	20	21	22	22	22
KYWA	3	4	8	18	22	25	25	26	26	26

3. Exsun Oil Company plans to invest $10 million during the current year. The investments options are three different projects, each of which can be developed at any of five different levels. The accompanying table shows the net present value or return in millions of dollars for each project at each available investment level. Find the optimal investment.

Project	Millions of dollars invested									
	1	2	3	4	5	6	7	8	9	10
1	7		7		14		18	20	21	
2				10	12			19	20	23
3						16	18	20	21	24

4. ABC Company follows an inventory policy in which inventory must be depleted before scheduling a new batch. A decision must be made as to how to schedule production for the next eight months. Assume that the demand per month is constant and is 500 units/month. The cost associated with the feasible batch sizes is given in the accompanying table. Use dynamic programming to find the least-cost production schedule.

Batch size	Cost in dollars
500	10,000
1000	12,000
1500	16,000
2000	22,000
2500	25,000
3000	31,000
3500	33,000

5. A truck can carry a maximum load of 10 tons, and three types of machinery, with weights of 1, 2.5, and 3 tons, respectively, must be loaded on it. The value to the company of each unit loaded is 0.10, 0.20 and 0.15, respectively. How many units of each type of machinery should be loaded in the truck so that total value is maximized?

6. Use dynamic programming to help a job-shop manager decide the optimal replacement policy for a new piece of equipment for the next five periods. The accompanying table shows the total costs, considering purchase price, operating costs (in thousands of dollars), and salvage value, when the machine is replaced at any given period and is kept for any additional periods. For instance, if the machine is replaced at period 2 and is kept for two additional periods, the company incurs a cost of $150.

Time of replacement	Additional periods machine is kept in use				
	1	2	3	4	5
1	90	140	155	160	175
2	95	150	215	235	
3	100	155	225		
4	125	160			
5	120				

7. Solve the following fixed-charge problem (see Section 7.5.4):

$$\text{Maximize} \quad x_1 + 3x_2 + g(x_3)$$
$$\text{subject to} \quad x_1 + 2x_2 + 3x_3 \leqslant 6$$
$$x_1 \qquad\quad + x_3 \leqslant 8$$
$$\text{where} \quad g(x_3) = \begin{cases} 0 & \text{if } x_3 = 0 \\ 4 + 2x_3 & \text{if } x_3 > 0 \end{cases}$$
$$x_1, x_2, x_3 \geqslant 0 \text{ and integer}$$

8. Solve the following knapsack problem:

Item	Weight (lb)	Value
1	1	1
2	2	3
3	3	2

The maximum load is 10 lbs.

9. Solve the following multidimensional knapsack problem:i

Item	Weight (lb)	Volume (ft^3)	Value
1	1	1	12
2	2	3	16
3	3	2	18
maximum	10	10	

10. Solve problem 13 of Chapter 2 (a linear programming problem) as a dynamic programming problem.

Nine

Search Techniques in Nonlinear Programming

9.1. Introduction

In Chapter 2 the general mathematical programming problem

$$\text{Optimize } f(X)$$

$$\text{subject to } g_i(X)\{\overset{\leq}{\underset{\geq}{=}}\}b_i \qquad i=1,2,\ldots,m$$

was introduced. Chapters 2 through 6 discussed the special case of this problem where f and g_i are all linear functions, enabling us to use linear programming to analyze the problem. Chapter 7 discussed the special integer case of linear programming and Chapter 8 discussed the special case in which f can be examined in stages. It would be nice if in this chapter a method for solving all mathematical programming problems of the general form shown above were developed. Unfortunately, once we leave the realm of linear functions there is no single way to solve a problem. Therefore, this chapter presents an insight into general mathematical programming problems and introduces different methods for solving different types of problems. Given a particular problem structure, an appropriate solution method can be chosen in such a way that it utilizes each special structure.

Essentially there are two approaches to solving a nonlinear programming problem. The first approach is to make use of the computational abilities of modern computers by evaluating a large number of points, determining which points are feasible, and then choosing the feasible point that yields the best value of the objective function. The alternative is to approach the problem from an analytical (calculus-based) viewpoint.

In Section 9.2 the computational methods for optimizing a function of one decision variable are discussed. Since there is only one variable, this approach is referred to as one-dimensional optimization. The techniques used are termed *line searches*, since in effect the line representing the x axis is searched in order to find the optimal point. The line searches of Section 9.2 not only are useful for problems with one variable but also form an integral part of the optimization techniques for the general n-variable nonlinear programming problems that are presented in Section 9.3. These methods are predicated on the speed and efficiency of modern computers. Other nonlinear programming methods are founded on several principles of calculus. In fact, the first optimization techniques ever used were simply results from calculus for unconstrained optimization problems. These results are known as *classical optimization techniques* and are presented in Section 9.4.

9.2. One-Dimensional Search Techniques

Before considering methods of solution it is imperative that we consider what types of problems can be solved and what is meant by "solution." Consider the following three functions and suppose that we are interested in finding the minimum of each of these functions:

$$f(x) = \begin{cases} 5-x & \text{if } 0 \leq x < 5 \\ x-4 & \text{if } 5 \leq x < 10 \end{cases}$$
$$g(x) = x \qquad 0 < x < 10$$
$$h(x) = 2x^3 - 9x^2 + 12x$$

The graph of each of these functions appears in Figure 9.1. It should be obvious from the graphs that there may be problems when finding the minimum of each of these functions.

As depicted in Figure 9.1 $f(\cdot)$ has no minimum due to the jump that occurs at $x=5$. The function $f(\cdot)$ is discontinuous and because of this discontinuity, as in this example, it is assumed that all functions examined henceforth are continuous. That is, the functions have no jumps and can be drawn without lifting the pencil from the paper.

The function $g(\cdot)$ entails a similar problem in that it has no minimum. This occurs due to the fact that the domain of the function is open at the ends because of the strict inequality constraints $x>0$ and $x<10$. For this reason only constraints of the type $\leq$, $=$, or $\geq$ are permitted.

The function $h(\cdot)$ has a different type of problem in that there is no minimum, but we might think that there is one at $x=2$, since the function is lower at $x=2$ than at any point "near" 2. Let us consider this in more detail.

Figure 9.2 exhibits an ill-behaved function. It is ill behaved because there are too many hills and valleys. It is possible that if the function is examined in the wrong region, then A will seem to be the minimum. Certainly the minimum over the entire region is at C, not A. However, since in a small

230

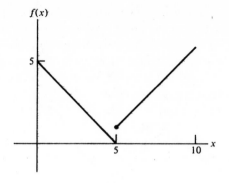

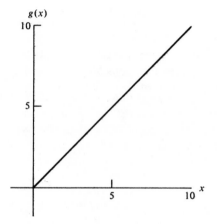

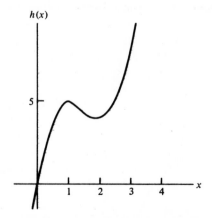

Figure 9.1. Ill-behaved Functions

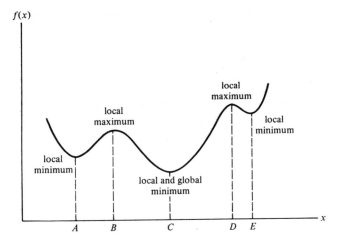

Figure 9.2. Local and Global Optima

region near *A* no point is lower than *A*, *A* is termed a *local minimum*. The point *C* is also a local minimum, but since it is the minimum over the entire region, it is also termed a *global minimum*. The key point is that when minimizing an ill-behaved function we may not find the global minimum but may find a local minimum. Similar definitions exist for *local maximum* and *global maximum*.

As mentioned earlier, a well-behaved function does not have hills and valleys. Therefore, it must appear as in Figure 9.3. From the left to the

Figure 9.3. A General Unimodal Function

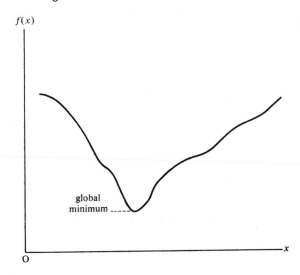

minimum it must decrease and then from the minimum to the right it must increase. Such a function is termed *unimodal*. With this in mind, we begin our study of the nonlinear optimization techniques. Consider the following example.

The Sweet Candy Company is beginning the production of a new dietetic candy bar. The production cost is 8.5¢ per bar. The marketing department has suggested that the weekly demand is given by $1000/p^2$ where p is the wholesale price at which the candy bar is sold. Sweet Candy Company is interested in finding the selling price that maximizes its total weekly profits.

Since weekly profit is simply the profit per bar times the number of bars sold, the profit function $Z(p)$ is given by

$$Z(p) = (p - 0.085)1000/p^2$$

The problem is to find the price, p, that maximizes $Z(p)$. Three different methods for finding the optimal price are presented in the next three subsections.

9.2.1. Brute Force Method

Suppose that the interest is in determining the maximum of $Z(p) = (p - 0.085)1000/p^2$ for $0.0 \le p \le 0.60$, but that due to computational cost the function $Z(p)$ can be evaluated only 11 times. That is, the profit for any 11 prices between 0 and 60¢ can be computed. A method that is obvious and quite reasonable is to examine $Z(p)$ for $p = 0, 0.06, 0.12, 0.18, \ldots, 0.60$. That is, begin at 0¢ and examine points spaced equally apart. A method of this type is termed a *brute force* method. These computations have been performed for the Sweet Candy problem, and the results can be seen in Table 9.1. Consider the information gained from the brute force search. From the table we would guess that the best price is near 18¢ and certainly between

Table 9.1. Brute Force Evaluation of Profit

Price p	Demand $1000/p^2$	Profit (loss)/bar $p - 0.085$	Total profit (loss) $(p - 0.085)1000/p^2$
0	∞	−0.085	−∞
0.06	277,000	−0.025	−6925
0.12	69,444	0.035	2431
0.18	30,864	0.095	2932
0.24	17,361	0.155	2691
0.30	11,111	0.215	2388
0.36	7,716	0.275	2122
0.42	5,668	0.335	1899
0.48	4,340	0.395	1714
0.54	3,429	0.455	1560
0.60	2,777	0.515	1431

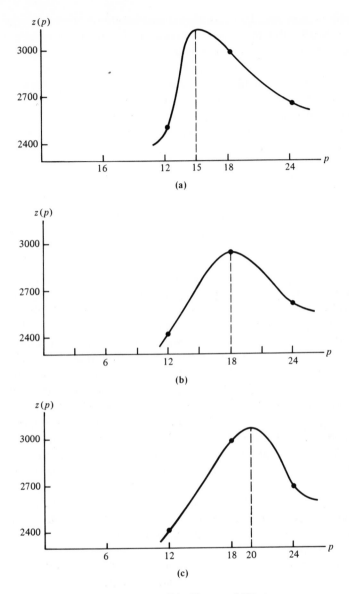

Figure 9.4. Possible Shapes of $Z(p)$

12¢ and 24¢. The reason we cannot say the maximum is at 18¢ can be seen in Figure 9.4. For the three values at 12¢, 18¢, and 24¢ it is possible that the function actually appears as any one of the cases in the figure. In Figure 9.4a, 15 is indicated as the maximum; in 9.4b, 18 is maximum; and in 9.4c, 20 is maximum. Actually the safest statement to be made is that "the optimal price lies between 12¢ and 24¢." (Since it is not known whether or not $Z(p)$ is unimodal, by optimal we mean local.) Of course, the denser the grid, the smaller this interval of uncertainty. Since each interval in the grid has a length of $L/(n-1)$ when n is the number of points searched and L is the size of the domain of the search, the interval of uncertainty is given by $2L/(n-1)$. We would like to develop a more efficient scheme using the same number of search points. Note that for the brute force method the points that are examined are determined in advance. The searches presented in the following two sections perform the computations sequentially making use of the current information.

9.2.2. Bisection Method

In this section a logically ordered procedure for reducing the interval of uncertainty is constructed. Consider again the function $Z(p)=(p-0.085)1000/p^2$; assume again that the initial interval of uncertainty is $[0, 0.6]$, and that as before 11 points will be examined. Let $x_1 = 0$ and $x_2 = 0.6$ and evaluate Z at these two boundary points, obtaining $Z(0) = -\infty$ and $Z(0.6) = 1431$. A logical point to consider next is the midpoint of the interval $[0, 0.6]$. Hence, let $x_3 = 0.3$ and find that $Z(0.3) = 2388$. The information available is that the profit begins at minus infinity, reaches at least \$2388, and is reduced to \$1431 at a price of 60¢. However, it is not known if profit increases to the right of 30¢ or not. The points that might help determine this are the midpoints of the intervals $[0, 0.30]$ and $[0.30, 0.60]$. Thus, search $x_4 = 15¢$ and $x_5 = 45¢$, respectively, and find that the associated profits are $Z(0.15) = 2888$ and $Z(0.45) = 1802$, respectively. Hence, the information appears as in Figure 9.5a. From the figure it can be determined that the maximum lies between 0 and 30¢. (Again, it is a global maximum if Z is well behaved, a local maximum otherwise.) Hence, after the initial two boundary point evaluations and the three midpoint evaluations, the interval of uncertainty is now one half the length of the original interval of uncertainty. We repeat this process on the new interval, $[0, 0.30]$. The points that must be evaluated are 0¢, 7.5¢, 15¢, 22.5¢ and 30¢. Notice that there already are values for 0¢, 15¢, and 30¢, so that actually only two new points need to be evaluated.

Letting $x_6 = 0.075$ and $x_7 = 0.225$, we determine that $Z(0.075) = -1777$ and $Z(0.225) = 2765$. The results are plotted in Figure 9.5b and we can see that the maximum must lie between 7.5¢ and 22.5¢. Again the interval of uncertainty has been reduced in length by one half. Continuing, we find that $x_8 = 0.1125$ and $x_9 = 0.1875$ and the respective profits are $Z(x_8) = 2172$ and $Z(x_9) = 2916$. Hence the new interval of uncertainty is between 15¢ and 22.5¢. One last bisection is made, and the evaluation at $x_{10} = 0.16875$ and $x_{11} = 0.20625$ is computed. The values are $Z(x_{10}) = 2941$ and $Z(x_{11}) = 2850$, respectively. The interval of uncertainty is now $[0.15, 0.1875]$, which has a length of 0.0375. Thus, evaluating the same number of points as with brute

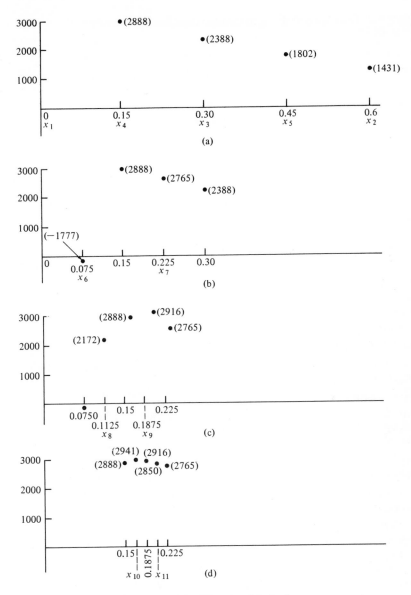

Figure 9.5. Bisection Method

force, we find that the final interval of uncertainty is shorter, 0.0375 compared with 0.12 in the brute force method. In general, the length of the interval is reduced by half for each pair of points added except for the first step, where five points are required. Thus, after n points have been examined the interval of uncertainty has a length that is not greater than $L/2^{(n-3)/2}$ for $n = 3, 5, 7, \ldots$. In Table 9.2 we see a comparison between

Table 9.2. Final Interval of Uncertainty for Bisection Method and Brute Force Method

Number of trials n	Bisection $1/2^{(n-3)/2}$	Brute force $2/(n-1)$
3	1	2/2
4	—	2/3
5	1/2	2/4
6	—	2/5
7	1/4	2/6
8	—	2/7
9	1/16	2/8
10	—	2/9
11	1/32	2/10
⋮		
25	1/2048	2/24
⋮		
101	2.10^{-15}	2/100

$2L/(n-1)$, which is the brute force length, and $L/2^{(n-3)/2}$, which is the bisection length. It can be seen that if seven or more points are examined, then the bisection method is superior. Typically, due to the speed of computers hundreds of points are examined and you can see from the table that for 101 points there is a ridiculously large difference in interval reduction.

One of the reasons for the superiority of the bisection method over the brute force method is that the bisection method takes advantage of the information obtained at each trial and chooses the next points to be evaluated according to this information. One similarity between the two methods should be noted: Each method divides the interval of uncertainty in a symmetrical manner.

In summary, the bisection method for optimizing a function defined on $[a, b]$ is as follows:

1. Set $x_1 = a$, $x_2 = a + (b-a)/4$, $x_3 = (a+b)/2$, $x_4 = a + 3(b-a)/4$, $x_5 = b$.
2. Compute $f(x_j), j = 1, 2, 3, 4, 5$.
3. Let j be such that $f(x_j) = \text{optimum } \{f(x_1), f(x_2), \ldots, f(x_5)\}$.
4. If $j = 2$, 3, or 4, then set $a = x_{j-1}, b = x_{j+1}$ and go to Step 7.
5. If $j = 1$, set $a = x_1, b = x_2$ and go to Step 7.
6. If $j = 5$, set $a = x_4, b = x_5$ and go to Step 7.
7. Set the interval of uncertainty equal to $L = b - a$. If L is less than the predetermined length, stop. Otherwise go to Step 1.

(For ease of explanation the subscripts in this summary are chosen differently from those in the example.

9.3.2. Fibonacci Search

A search that is similar to the bisection method but reduces the interval of uncertainty even more is the Fibonacci search. Again, the goal is to have an interval of uncertainty that is as small as possible for a given number of trials. Suppose that the interval $[a, b]$ of length $L = b - a$ is to be searched using n search points. Let F_k be the kth element of the Fibonacci sequence. The Fibonacci sequence is given by the equation $F_k = F_{k-1} + F_{k-2}$ with $F_0 = F_1 = 1$. Hence, $F_2 = F_1 + F_0 = 1 + 1 = 2$, $F_3 = F_2 + F_1 = 2 + 1 = 3$, $F_4 = 3 + 2 = 5$, $F_5 = 5 + 3 = 8$, etc. (The Fibonacci sequence was originally developed as a model of the reproduction of rabbits; see Gardner, 1979.) Table 9.3 presents the first twenty numbers of the sequence.

For example, suppose that $n = 6$, the function is $Z(p) = 1000(p - 0.085)/p^2$, and the interval to be searched is $[a, b] = [0, 0.65]$. (The interval has been expanded from 60ε in the previous examples to 65ε in order to avoid round-off difficulties. Begin by searching the function at the two points that are located $(F_{n-1}/F_n)L$ from a and b. In this case since $n = 6$, $F_{n-1}/F_n = F_5/F_6$, which from Table 9.3 is equal to $8/13$. Hence, the two points are $x_1 = 0 + 8/13(0.65) = 0.40$ and $x_2 = 0.65 - 8/13(0.65) = 0.25$. Notice that as with the previous two searches the interval is divided symmetrically. The values found are $Z(0.40) = 1969$ and $Z(0.25) = 2640$, and are shown in Figure 9.6a. Since $Z(0.40) < Z(0.25)$, the new interval of uncertainty is $[0, 0.40]$ and this interval should be divided symmetrically. Hence, let $x_3 = 0 + (0.40 - 0.25) = 0.15$. Since $Z(0.15) = 2888 > Z(0.25)$ the new interval of uncertainty is $[0, 0.25]$, which again must be divided symmetrically. Thus, let $x_4 = 0 + (0.25 - 0.15) = 0.10$ and find that $Z(0.10) = 1500$. The new interval of uncertainty is $[0.10, 0.25]$ and the symmetrical partition yields $x_5 = 0.10 + (0.25 - 0.15) = 0.20$ with $Z(0.20) = 2875$. The interval of uncertainty is now $[0.10, 0.20]$ and partitioning again yields $x_6 = 0.10 + (0.20 - 0.15) = 0.15$. However, 0.15 is a point that has already been searched ($x_3 = 0.15$). Thus to gain another division in the interval, search slightly to the left or to the right of 0.15. Setting $x_6 = 0.151$, we find that $Z(0.151) = 2894$ and the final interval of uncertainty is $[0.15, 0.20]$. (This interval is larger than in the brute force or bisection examples. However, recall that the initial interval of uncertainty was larger and that 6 points are examined rather than 11 points.) Notice that the final interval of uncertainty has a length of

Table 9.3. The Fibonacci Sequence

n	0	1	2	3	4	5	6	7	8	9	10	11	12	13	14	15	16	17	18	19	20
F_n	1	1	2	3	5	8	13	21	34	55	89	144	233	377	610	987	1597	2584	4181	6765	10,946

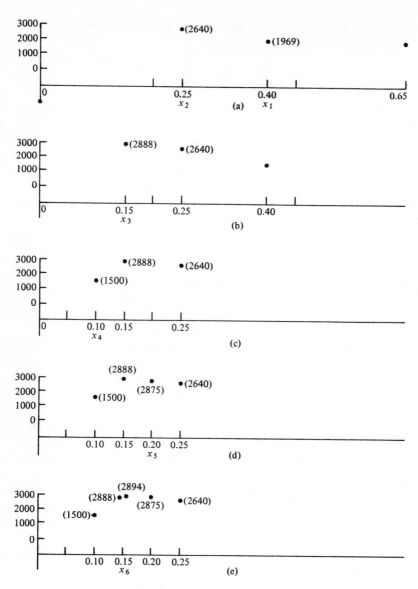

Figure 9.6. Fibonacci Search

$(1/13)L$ or $(1/F_n)L$ where n is the number of points searched. The Fibonacci search generates a final interval of uncertainty of $(1/F_n)L$. Table 9.4 contains a comparison of the Fibonacci search and the bisection method. It can be seen from the table that if the number of points examined is three or more, then the Fibonacci method is superior to the bisection method (which is, as mentioned before, superior to brute force). There is

Table 9.4. Final Interval of Uncertainty for Fibonacci and Bisection Methods

n	Fibonacci $1/F_n$	Bisection $1/2^{(n-3)/2}$
3	1/3	1
4	1/5	—
5	1/8	1/2
6	1/13	—
7	1/21	1/4
8	1/34	—
9	1/55	1/8
10	1/89	—
11	1/144	1/16
12	1/233	—
13	1/377	1/32
14	1/610	—
15	1/987	1/64
16	1/1597	—
17	1/2584	1/128
18	1/4181	—
19	1/6765	1/256

one disadvantage to the Fibonacci search in that the number of trials, n, must be known from the outset and cannot be changed.

In summary, three of the possible search techniques that can be used when optimizing a function of one variable have been presented. The search techniques vary in both their efficiency and their ease of use. The easiest to apply (program) is the brute force method. Also, the trials are independent of one another, which is not true for the bisection and Fibonacci methods.

Table 9.5. Comparison of Search Techniques

	Brute Force	Bisection	Fibonacci
Final interval of uncertainty	$\dfrac{2L}{n-1}$	$\dfrac{L}{2^{(n-3)/2}}$	$\dfrac{L}{F_n}$
Addition of new points after completion of search	yes	yes	no
Trial points determined in advance	yes	no	no
Comments	Most flexible method; easiest to program	Number of trials can vary. Location of nth trial depends on results of $(n-1)$st trial.	Most efficient, least flexible. Location of nth trial depends on results of $(n-1)$st trial.

Brute force yields the largest interval of uncertainty. The bisection method improves the interval of uncertainty but increases computation costs and programming time and this is even more true of the Fibonacci search. The Fibonacci method has another disadvantage in that n is fixed and cannot be changed. The information on the three searches is summarized in Table 9.5.

9.3. Multidimensional Search Techniques

The one-dimensional search techniques are rather simple in that movement is always to the left or right. In two-dimensional problems movement can be up or down as well as from the left to the right, and in the n-dimensional problem there are n choices, one for each dimension. In this section two methods for solving n-dimensional problems are presented.

9.3.1. Grid Search

In this section, the brute force method is extended to more than one dimension. Figure 9.7 includes one-, two-, and three-dimensional diagrams of the points that might be examined in these respective cases. It can be seen that *each* axis is divided into a number of equally spaced intervals and that the search is along a line, rectangle, or box, depending on the dimension. Notice that the two-dimensional version forms a grid; hence the name of the search. The total number of points searched for an n-dimensional problem is $(m_1 + 1)(m_2 + 1) \cdots (m_n + 1)$ where m_j is the number of intervals along axis j.

Even if the function is well behaved, there still are problems in determining the exact optimal value. In one dimension the optimum is within one length left or right of the minimum that is found. In n dimensions the optimum is within one length in every direction. For example, in the two-dimensional problem illustrated in Figure 9.8, all that can be said is that the optimum is within the square $abcd$ if X is the minimum value of all of the grid points examined.

9.3.2. Feasible Direction Method

In one dimension brute force methods are inferior to binary searches or Fibonacci searches. The same is true for the grid search. That is, more efficient methods can be developed. The general idea is to start at an arbitrary point, choose a direction in which to search, and then examine a (one-dimensional) line using the techniques of Section 9.2. This concept is presented in Figure 9.9. In the figure, X_1 is the starting point and d_1 is the direction in which the search is made. The best value along this line is, say, X_2. A new direction d_2 is chosen and line 2 is examined, resulting in the best value at X_3. This process is repeated as many times as desired. Each point in

(a) one–dimensional grid

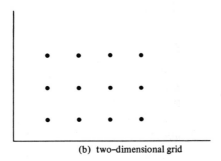

(b) two–dimensional grid

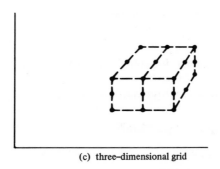

(c) three–dimensional grid

Figure 9.7. One-, Two- and Three-Dimensional Grid Patterns

Figure 9.8. Multidimensional Interval of Uncertainty

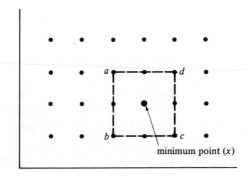

minimum point (x)

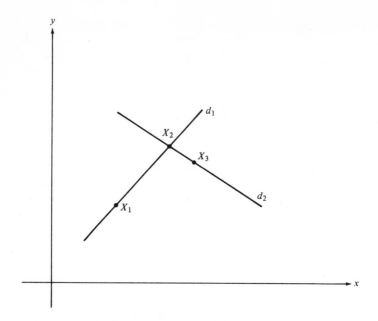

Figure 9.9. Feasible Direction Method

the sequence should improve the solution (as is true with the simplex method).

Consider the unconstrained problem

$$\text{Minimize} \quad x^2 - 6x + y^2 - 8y + xy$$

and suppose that the initial guess at a solution is $X_1 = (x, y) = (1,2)$. Note that the functional value at X_1 is $1^2 - 6 \cdot 1 + 2^2 - 8 \cdot 2 + 1 \cdot 2 = 1 - 6 + 4 - 16 + 2 = -15$. Suppose that the first direction chosen is arbitrarily taken to be $d_1 = (1,4)$. That is, the new point, X_2, will be given by $X_2 = X_1 + r(1,4)$ for some value of r. The problem generated is

Minimize $f(x, y) = f(1 + 1r, 2 + 4r)$

$$= (1 + 1r)^2 - 6(1 + 1r) + (2 + 4r)^2 - 8(2 + 4r) + (1 + 1r)(2 + 4r)$$

$$= 1 + 2r + r^2 - 6 - 6r + 4 + 16r + 16r^2 - 16 - 32r + 2 + 6r + 4r^2$$

$$= 21r^2 - 14r - 15 = f(r)$$

The key point is that the foregoing is a function of only one variable, namely r, and can be optimized by using the techniques of the previous section. A line search reveals that $f(r)$ is minimized at $r = \frac{1}{3}$. The new point X_2 is given by $X_2 = (1,2) + (1,4)/3 = (4/3, 10/3)$ and the functional value at X_2 is -17.67. Hence, the objective function has improved.

It happens that $(4/3, 10/3)$ is optimal (see Problem 16); hence the search is finished. This, of course, was lucky in that the right direction was initially chosen.

In summary, the general approach is to split the solution method into two steps:

1. Find the direction in which to search.
2. Find the optimal point along this direction.

In all cases Step 2 may be executed by the line searches previously presented. Hence, the only problem is determining the direction in which to search.

One simple solution is always to search along lines that are parallel to the axes. For a three-dimensional problem the first direction would be, $(1,0,0)$, the second direction would be $(0,1,0)$, the third direction would be $(0,0,1)$, the fourth direction would be, to repeat, $(1,0,0)$, and so on. The method that consists in using these directions is termed the *cyclical coordinates* method.

For the example above let us use the same starting point, $X_1=(1,2)$. The first direction is $(1,0)$; hence the problem is

$$\text{Minimize}_r f(x, y)=f(1+1r,2+0r)$$

$$=(1+1r)^2-6(1+1r)+(2)^2-8(2)+(1+1r)(2)$$
$$=1+2r+r^2-6-6r+4-16+2+2r$$
$$=r^2-2r-15=f(r)$$

Again, the problem is a simple one-dimensional search and the optimal value of r is 1. Thus $X_2=(1,2)+1(1,0)=(2,2)$ and $f(2,2)$ is equal to

$$2^2-6\cdot2+2^2-8\cdot2+2\cdot2=4-12+4-16+4=-16$$

Hence, the objective function has indeed improved. The next point will be $X_3=(2,2)+r(0,1)$; to find it, we solve the problem

$$\text{Minimize}_r f(x, y)=f(2+0r,2+1r)$$

$$=2^2-6\cdot2+(2+r)^2-8(2+r)+2(2+r)$$
$$=4-12+4+4r+r^2-16-8r+4+2r$$
$$=r^2-2r-16=f(r)$$

Again the optimizing value of r is found by a line search and is equal to 1. Hence, the new point is $X_3=(2,3)$, which yields a functional value of -17 and the new problem, using the direction $(1,0)$, is

$$\text{Minimize } f(x, y)=f(2+1r,3+0r)$$

$$=(2+r)^2-6(2+r)+3^2-8(3)+(2+r)3$$
$$=4+4r+r^2-12-6r+9-24+6+3r$$
$$=r^2+r-17=f(r)$$

The optimizing value is given by $r = -0.5$, which means that $X_4 = (1.5, 3)$ and generates a functional value of -17.25. Again, the function has been improved. It is possible to continue in this fashion for as long as desired, each time improving the objective function.

Constrained Problems. So far, it has been assumed that any direction can be chosen and that movement may be as far as desired along a line. In a constrained problem this may not be true.

Consider the feasible region given by the constraints $4x + y \leqslant 20$ and $x + 4y \leqslant 20$, as shown in Figure 9.10. (The nonnegativity restrictions are assumed.) Suppose that the current solution is some point X (strictly) inside of the feasible region. For example, X might be the point $(3, 3)$. Note that from the point $(3, 3)$ we can travel in any direction for some distance. Of course, the amount we can move along any line depends on the direction chosen. If the current point is inside the feasible region, that is, the point does not touch a constraint line, any direction can be chosen. After choosing the direction, find the length of the line to be searched. For example, if the current point is $(3, 3)$ and the direction is $(1, 1)$, the line

Figure 9.10. Constrained Choice of Direction

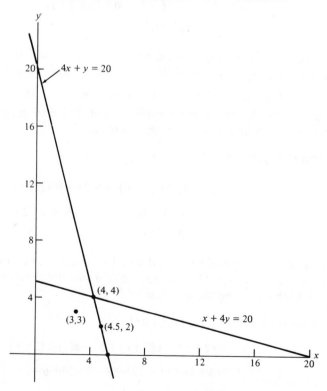

cannot go beyond $(4,4)$. After finding the end of the line segment, perform a line search on the segment to generate the next point.

Next, consider the case where the current point is on the boundary of the feasible region. In this case, every direction is not necessarily feasible. For example, if $X=(x, y)=(5,0)$, then obviously the direction cannot increase in x or decrease in y. When on the boundary, it is necessary to distinguish between the active and inactive constraints and concern ourselves only with active constraints. Suppose the current point is $(4.5,2)$. Then it is necessary to stay to the left of the constraint $4x+y\leqslant 20$. Let the direction be denoted by (d_1, d_2) with a step size of r. Then it must be guaranteed that

$$4(x+rd_1)+(y+rd_2)\leqslant 20$$

given that $4x+y=20$ (since the current point is on the boundary). In other words, it must be true that

$$4rd_1+rd_2\leqslant 0$$

or since $r>0$, $4d_1+d_2\leqslant 0$. Any direction that satisfies the last inequality is a feasible direction and we can proceed as before.

Now suppose that the current point is $(4,4)$. In this case there are two active constraints. As before it must hold that $4d_1+d_2\leqslant 0$ in order to satisfy the first constraint. The other constraint is satisfied if $\frac{1}{4}d_1+d_2\leqslant 0$. Thus the problem is

$$\text{find} \qquad (d_1, d_2)$$
$$\text{subject to} \quad 4d_1+d_2\leqslant 0$$
$$\tfrac{1}{4}d_1+d_2\leqslant 0$$

Notice that d_1 and d_2 are unconstrained. For this reason a boundary condition must be added that ensures that d_1 and d_2 are finite. The easiest condition to add is

$$|d_1|+|d_2|\leqslant 1$$

In other words, to find a feasible direction we need only solve a linear programming problem (see Problem 19).

9.4. Classical Optimization Techniques*

9.4.1. Unconstrained Optimization Techniques

Consider again the Sweet Candy problem. Let us use calculus to examine the function $Z(p)$ where

$$Z(p)=1000/p-85/p^2$$

---

*The material in this section requires knowledge of differential calculus and may be skipped without loss of continuity.

It follows that the first derivative, $Z'(p)$, is given by

$$Z'(p) = -1000/p^2 + 170/p^3$$

Setting the first derivative equal to zero and solving for p yields

$$0 = -1000p + 170$$
$$p = 170/1000 = 0.17$$

Now let p be any number less than 0.17; then $Z'(p)$ is greater than zero, whereas if p is any number greater than 0.17, $Z'(p)$ is less than zero. Since the sign of the first derivative determines whether the function is increasing (first derivative positive) or decreasing (first derivative negative) some conclusions about $Z(p)$ can be drawn. For any price less than 17¢ profits are increasing as the price increases, while for any price greater than 17¢, profits are decreasing as price increases. At a price of 17¢ the rate of change of total profits is zero. Thus 17¢ must be the optimal price. That is, 17¢ is at least a local maximum.

From this example there are some general rules that can be determined for functions that have a derivative at every point. Note that since the first derivative yields information about the trend of a function, then if at some point x the first derivative is positive, a point just to the right of x increases the function. Similarly, if the first derivative is negative at x, a point just to the left of x increases the function. Therefore, if the first derivative is positive or negative at some point x, this point cannot possibly maximize the function. The only value left for the first derivative is zero. Hence the following must be true.

Rule 1A. If x^* is a local maximum of $f(x)$, then $f'(x^*) = 0$.

Similarly, when minimizing a function and the derivative is either positive or negative at some point x the value of the function can be reduced by moving left or right, respectively. This yields

Rule 1B. If x^* is a local minimum of $f(x)$, then $f'(x^*) = 0$.

Notice that the rules do not say that if $f'(x) = 0$, then x minimizes or maximizes $f(x)$. For this reason $f'(x) = 0$ is called a necessary condition for optimality. Also, a point x that satisfies $f'(x) = 0$ is called a *stationary point*. In order for x to be optimal it is necessary that this condition ($f'(x) = 0$) be met, but the condition is not sufficient. That Rule 1A is only necessary and not sufficient is best exemplified by the following problem:

$$\text{Maximize } f(x) = (x-4)^3$$

Notice that $f'(x) = 3(x-4)^2$. The value of the function is 0 when x is 4 but this is not a maximum since $x = 5$ yields the better value, 1. In fact, since

$f'(x)>0$ if $x>4$, any number larger than 4 yields a larger value for $f(x)$. Since it is necessary for the first derivative to be zero and this does not yield a maximum, it must be concluded that the function has no maximum. Alternatively, we might say that the maximum occurs at $x=+\infty$. The point $x=5$ is an inflection point rather than a maximum or minimum. The function is also infinitely small at $x=-\infty$.

Also notice that Rule 1A or 1B applies only to local optima. As a second example of the deficiency of Rule 1A, consider the problem

$$\text{Maximize } g(x)=x^3-2x^2+x+5$$

Setting the first derivative to zero yields

$$3x^2-4x+1=0$$

or

$$x=\frac{4\pm\sqrt{16-12}}{6}=\frac{2}{3}\pm\frac{1}{3}$$

Hence the only candidates for maximizing points are $x=1/3$ and $x=1$. Since $g(1/3)=(1/3)^3-2(1/3)^2+(1/3)+5=1/27-2/9+1/3+5=139/27$ and $g(1)=1-2+1+5=5$, $x=1/3$ is the only value that can maximize $g(x)$. Yet $g(2)=2^3-2\cdot2^2+2+5=7$, which is greater than $g(1/3)$, so $x=1/3$ does not maximize $g(x)$. Furthermore, notice that $g'(x)>0$ for $x>1$; thus $g(x)$ is increasing on $(1,+\infty)$. Again we must conclude that $g(x)$ has no global maximum.

Although $x=\frac{1}{3}$ is not a maximum over the entire x axis, it is a special point in that no point near it yields a higher value. Thus a local maximum has in fact been found.

Let us now develop the sufficient conditions for finding whether a stationary point is a local minimum, a local maximum, or neither. From the previous analysis it should be clear that for a point x^* to be a local maximum, the function must increase to x^* and decrease beyond x^*. That is, near x^* the function should appear to be unimodal. Thus for any point x just to the left of x^*, $f'(x)$ must be positive and for any point x just to the right of x^*, $f'(x)$ must be negative. Of course, $f'(x^*)$ must still be zero. This can be expressed as follows. Let ε be some very small positive number. Then a sufficient condition for a stationary point x^* to be a local maximum is given by

Rule 2A. x^* is a local maximum if $f'(x)>0$ for $x^*-\varepsilon<x<x^*$ and $f'(x)<0$ for $x^*<x<x^*+\varepsilon$.

Unfortunately this rule is usually difficult to check. Now, suppose it is true that for some point x^*, $f'(x^*)=0$ and that on the right of x^*, $f'(x)$ is decreasing. That is, $f'(x)<0$ for $x>x^*$. Furthermore, suppose that for any $x<x^*$, $f'(x)$ is increasing. Then $f'(x)>0$ for $x<x^*$. In other words, if $f'(x)$

is always decreasing and $f'(x^*)=0$, then $f(x)$ satisfies Rule 2A. To say that $f'(x)$ is decreasing is the same as saying that the derivative of $f'(x)$ is negative or in other words that $f''(x^*)<0$. This leads us to a sufficient condition (which is stricter than Rule 2A).

Rule 2B. If $f'(x^*)=0$ and $f''(x^*)<0$, then x^* is a local maximum of $f(x)$.

Similarly the analogous conditions for minimization problems are

Rule 2C. If $f'(x^*)=0$ and $f''(x^*)>0$, then x^* is a local minimum of $f(x)$.

Let us check the profit function for the Sweet Candy problem.

$$Z'(p)=-1000/p^2+170/p^3$$

and solving $-1000/p^2+170/p^3=0$ yields $p=0.17$

$$Z''(p)=2000/p^3-510/p^4$$

and since for $p=0.17$

$$Z''(0.17)<0$$

$p=0.17$ is a local maximum.

Let us look at a second example—the function examined before, which is given by $f(x)=(x-4)^3$. The derivatives are

$$f'(x)=3(x-4)^2$$

and

$$f''(x)=6(x-4)$$

The first derivative is zero at $x=4$. Since the second derivative is neither strictly positive nor strictly negative at $x=4$ (but the second derivative is zero), we cannot conclude that the necessary conditions are satisfied. (Other means exist and should be used for this determination.) Finally, consider the function $g(x)=x^3-2x^2+x+5$. The derivatives are

$$g'(x)=3x^2-4x+1$$

and

$$g''(x)=6x-4$$

Recall that $g'(x)=0$ at $x=1$ and $x=\frac{1}{3}$. Note that at $x=1$, $g''(x)=2$. Hence $x=1$ satisfies the conditions sufficient for a local minimum. For $x=\frac{1}{3}$, $g''(x)=-2$. Thus $x=\frac{1}{3}$ is a local maximum.

The last concern is determining when the local optimum is a global optimum. Again, we would like to check to see if the function is unimodal. This is difficult, however, where there are conditions that are stricter but easier to use.

Rule 3A. If $f'(x^*)=0$ and $f''(x)\leqslant0$ for all x, then x^* is a global maximum.

Rule 3B. If $f'(x^*)=0$ and $f''(x)\geqslant0$ for all x, then x^* is a global minimum.

A function that satisfies $f''(x)\leqslant0$ for all x is termed concave. Similarly, a function that satisfies $f''(x)\geqslant0$ for all x is termed convex. Hence Rules 3A and 3B may be expressed as

Rule 3C. For a concave (convex) function a local maximum (minimum) is a global maximum (minimum).

More will be said about convex functions momentarily.

Rules 2 and 3 are sufficient conditions but not necessary. For example, the Sweet Candy problem does not satisfy Rule 3A, yet 17¢ is a globally optimal price. Let us consider two more examples.

Example 1. Minimize $f(Q)=DK/Q+hQ/2$.

$$f'(Q)=-DK/Q^2+h/2$$

and

$$f''(Q)=2DK/Q^3$$

Thus if D and K have the same sign, $f(Q)$ is convex on $(0,+\infty)$ and concave on $(-\infty,0)$ and not defined for $Q=0$. The roots of $f'(Q)=0$ are $\pm\sqrt{2DK/h}$ and hence the local minimum occurs at $\sqrt{2DK/h}$ and the local maximum occurs at $-\sqrt{2DK/h}$. The global minimum is at $-\infty$ and the global maximum is at $+\infty$.

Example 2. Minimize $x^4+35x^2-50x+24$.

$$f'(x)=4x^3+70x-50$$

$$f''(x)=12x^2+70$$

Since $f''(x)$ is positive for all x, $f(x)$ is convex and the local minima are global minima. Solving $f'(x)=0$ yields the global minima; however, this is no easy task. A binary search method for solving this problem will be given shortly.

In summary, the necessary conditions for local optimality are the following:

Rule 1A. If x^* is a local maximum of $f(x)$, then $f'(x^*)=0$ if it exists.

Rule 1B. If x^* is a local minimum of $f(x)$, then $f'(x^*)=0$ if it exists.

Sufficient conditions for local optimality are

> **Rule 2A.** x^* is a local maximum if $f'(x)>0$ for $x^*-\varepsilon<x<x^*$ and $f'(x)<0$ for $x^*<x<x^*+\varepsilon$ and $\varepsilon>0$.
>
> **Rule 2B.** If $f'(x^*)=0$ and $f''(x^*)<0$, then x^* is a local maximum of $f(x)$.
>
> **Rule 2C.** If $f'(x^*)=0$ and $f''(x^*)>0$, then x^* is a local minimum of $f(x)$.

Sufficient conditions for global optimality are

> **Rule 3A.** If $f'(x^*)=0$ and $f''(x)\leqslant0$ for all x, then x^* is a global maximum.
>
> **Rule 3B.** If $f'(x^*)=0$ and $f''(x)\geqslant0$ for all x, then x^* is a global minimum.

Up to this point only functions of one variable have been considered. However, the extension to functions of more than one variable is very straightforward. Consider the function $f(x_1, x_2,\ldots, x_n)$. Then the three rules corresponding to Rules 1, 2, and 3 using partial derivatives $\partial/\partial x_j$; are

Rule 4 (necessary). If $(x_1^*, x_2^*,\ldots, x_n^*)$ optimizes $f(x_1, x_2,\ldots, x_n)$ then

$$\frac{\partial}{\partial x_j} f(x_1^*, x_2^*,\ldots, x_n^*)=0 \qquad \text{for } j=1,2,\ldots, n$$

Letting $x^*=(x_1^*, x_2^*,\ldots, x_n^*)$, we can write Rule 4 as

$$\nabla f(x^*)=\left(\frac{\partial}{\partial x_1} f(x^*), \frac{\partial}{\partial x_2} f(x^*),\ldots, \frac{\partial}{\partial x_n} f(x^*)\right)=(0,0,\ldots,0)$$

The term $\nabla f(x^*)$ is the gradient of f evaluated at the point x^* and is the vector of partial derivatives with respect to each of the n variables. Note that if n is one (the function is one-dimensional), then the gradient of f is precisely the derivative of f. Hence, the general necessary condition for optimality is that the gradient must be identically equal to zero.

Rule 5. (local minimum (maximum)—sufficient). If $f(x_1, x_2,\ldots, x_n)$ is convex (concave) at $(x_1^*, x_2^*,\ldots, x_n^*)$ and the partial derivatives are zero, then $(x_1^*, x_2^*,\ldots, x_n^*)$ is a local minimum (maximum).

Rule 6. (global minimum (maximum)—necessary and sufficient).
If $f(x_1, x_2,\ldots, x_n)$ is convex (concave) everywhere and if $x^*=(x_1, x_2,\ldots, x_n)$ satisfies Rule 4, then x^* is a global minimum (maximum).

There are several definitions of convexity. For one dimensional functions we have the definition given immediately after Rule 3B. However, for multidimensional functions the one easiest to check is usually as follows: $f(X)$ is convex if for any X, Y, and λ, $0\leqslant\lambda\leqslant1$, $f(\lambda X+(1-\lambda)Y)\leqslant$

$\lambda f(X)+(1-\lambda)f(Y)$. Geometrically, a line drawn between two points on the function lies above the function.

Consider the following example of Rule 6.

$$\text{Minimize } f(x, y)=x^2-6x+y^2-8y+25$$

Then the necessary condition is

$$\frac{\partial}{\partial x}f(x, y)=2x-6=0$$

and

$$\frac{\partial}{\partial y}f(x, y)=2y-8=0$$

Hence, the point $(x=3, y=4)$ is the only candidate for a global optimum. Furthermore, $f(x, y)$ is convex (see Problem 15 at the end of this chapter), so $(3,4)$ is the global minimum.

Searching with derivatives. Although the search techniques of Section 9.2 can be used without derivatives, it is worthwhile to consider the bisection method when applied in conjunction with a derivative. Suppose that the objective is to optimize a function $f(x)$ on the range $[a, b]$. Also, assume that the derivative exists and is given by $f'(x)$. Rather than finding the points that optimize $f(x)$ we can search for the points that satisfy $f'(x)=0$. Thus the optimization problem is converted to a problem of finding the root(s) of the equation $g(x)=0$ where in this case $g(x)=f'(x)$.

Now suppose that $g(a)<0$ and $g(b)>0$. Then somewhere between $x_1=a$ and $x_2=b$, $g(\cdot)$ crosses zero. (See Figure 9.11.) The goal is to find the point of crossing. Suppose that we examine the midpoint $x_3=(a+b)/2$; then either $g(x_3)<0$, $g(x_3)=0$, or $g(x_3)>0$. If $g(x_3)=0$, then the problem is, of course, completed. If $g(x_3)<0$, then somewhere between x_3 and b, $g(x)$ crosses zero. If $g(x_3)>0$, then somewhere between a and x_3, $g(x)$ crosses zero. In any case, either the problem is solved or the interval of uncertainty is reduced by half. We then continue with the new problem defined on either $[a, x_3]$ or $[x_3, b]$.

As an example, consider the problem of maximizing $Z(p)=1000(p-0.085)/p^2$. The derivative is $Z'(p)=-1000/p^2+170/p^3$. If the initial interval of uncertainty is $[0,0.60]$, then $Z'(0)>0$, $Z'(0.60)<0$. For $p=0.30$, $Z'(0.30)<0$; hence Z' crosses zero between 0 and 0.30. Now for $p=0.15$, $Z'(0.15)>0$; hence the new interval of uncertainty is $[0.15,0.30]$. Since $Z'(0.225)$ is negative, the interval of uncertainty is $[0.15,0.225]$.

The process may be continued as much as desired. For each derivative examined, the interval of uncertainty is reduced in half. Hence using the derivative makes the bisection method work twice as fast as when using the bisection method on the original function, since the original method requires two evaluations in order to cut the interval of uncertainty in half (see Table 9.5).

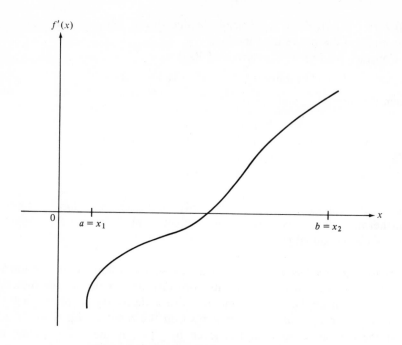

Figure 9.11. Binary Search with Derivatives

9.4.2. Constrained Optimization

Let us extend our analysis to problems that include constraints. As indicated in the introduction to this chapter, different methods are used for different types of nonlinear programming problems. The first type of nonlinear programming structure that is considered is that of a nonlinear objective function of two variables and one constraint of the equality form. The method of analysis for this problem is termed the *method of Lagrange multipliers.*

Consider the two-dimensional problem

$$\text{Minimize } (x_1 - 3)^2 + (x_2 - 4)^2$$

$$\text{subject to } 2x_1 + x_2 = 7$$

The unconstrained minimum occurs at $(3,4)$ and is equal to zero. One way to find the constrained minimum is by substitution. That is, since $x_2 = 7 - 2x_1$, the problem is to minimize $f(x_1)$ where

$$
\begin{aligned}
f(x_1) &= (x_1 - 3)^2 + (7 - 2x_1 - 4)^2 \\
       &= (x_1 - 3)^2 + (3 - 2x_1)^2 \\
       &= x_1^2 - 6x_1 + 9 + 9 - 12x_1 + 4x_1^2 \\
       &= 5x_1^2 - 18x_1 + 18
\end{aligned}
$$

Since $f'(x_1)=10x_1-18$ and $f''(x_1)=10>0$, the global minimum occurs at $x_1=18/10$ and $x_2=7-2(1.8)=3.4$. The optimal value of the objective function is $(1.8-3)^2+(3.4-4)^2=1.8$. Thus the constraint has increased the objective function by 1.8.

Since it is relatively easy to solve the problem by using the necessary and sufficient conditions developed in Section 9.3, let us consider a general method for doing this. The mean is the construction of a related problem that is unconstrained. The manner in which the unconstrained problem is treated is to incorporate the constraint into the objective function.

Consider the following function:

$$L(x_1, x_2, \lambda)=(x_1-3)^2+(x_2-4)^2+\lambda(2x_1+x_2-7)$$

Note that L is a function of the original two variables and a new variable λ. Also, L is the sum of the original objective function and a multiplier of the original constraint. Suppose that the problem considered is

$$\underset{\text{over } x_1, x_2, \lambda}{\text{Minimize}} \; L(x_1, x_2, \lambda)$$

The necessary conditions in order to minimize $L(x_1, x_2, \lambda)$ are that the three partial derivatives be equal to zero. Thus, it must be that

1. $\dfrac{\partial}{\partial x_1} L(x_1, x_2, \lambda)=2(x_1-3)+2\lambda=0$

2. $\dfrac{\partial}{\partial x_2} L(x_1, x_2, \lambda)=2(x_2-4)+\lambda=0$

3. $\dfrac{\partial}{\partial \lambda} L(x_1, x_2, \lambda)=2x_1+x_2-7=0$

Condition 3 simply states that $2x_1+x_2=7$ if (x_1, x_2, λ) is an optimal solution, or that (x_1, x_2) is feasible for the original problem. Also, if (x_1, x_2, λ) optimizes $L(x_1, x_2, \lambda)$, then the term $\lambda(2x_1+x_2-7)$ is equal to $\lambda \cdot 0=0$ and thus $L(x_1, x_2, \lambda)=(x_1-3)^2+(x_2-4)^2$, which is the objective function for the original problem. Hence, if (x_1, x_2, λ) optimizes $L(x_1, x_2, \lambda)$, then (x_1, x_2) is *feasible* and *optimal* for the original problem. Now, conditions 1, 2, and 3 are simply three equations with three unknowns, x_1, x_2, λ; hence, these equations can be solved, yielding $x_1=1.8$, $x_2=3.4$, and $\lambda=1.2$ as a solution. Thus $x_1=1.8, x_2=3.4$ is the optimal solution to the original problem. In general, for a problem with two variables, x_1 and x_2, an objective function $f(x_1, x_2)$ and one constraint $g(x_1, x_2) \leqslant b$, create the Lagrangian function

$$L(x_1, x_2, \lambda)=f(x_1, x_2)+\lambda[g(x_1, x_2)-b]$$

and set the three partial derivatives to zero in order to solve the problem.

This concept can be extended to any number of variables and constraints. Suppose that the problem is a general nonlinear programming problem given by

Minimize $\qquad f(X)$

subject to $\qquad g_i(X)=b_i \qquad i=1,2,\ldots,m$

$X=(x_1,x_2,\ldots,x_n)$

Define the Lagrangian as

$$L(X,\Lambda)=f(X)+\sum_{i=1}^{m}\lambda_i[g_i(X)-b_i]$$

$$=f(X)+\lambda_1[g_1(X)-b_1]+\lambda_2[g_2(X)-b_2]+\cdots+\lambda_m[g_m(X)-b_m]$$

where $\Lambda=(\lambda_1,\lambda_2,\ldots,\lambda_m)$. The necessary conditions for the unconstrained problem

Minimize $L(X,\Lambda)$

over all $X=(x_1,x_2,\ldots,x_n)$

all $\Lambda=(\lambda_1,\lambda_2,\ldots,\lambda_m)$

are

$$\frac{\partial}{\partial x_j}L(X,\Lambda)=\frac{\partial}{\partial x_j}f(X)+\sum_{i=1}^{m}\lambda_i\frac{\partial}{\partial x_j}g_i(X)=0 \qquad j=1,2,\ldots,n$$

$$\frac{\partial}{\partial\lambda_i}L(X,\Lambda)=g_i(X)-b_i=0 \qquad i=1,2,\ldots,m$$

Hence there are $n+m$ equations with $n+m$ unknowns and the solution to these equations yields a stationary point for the original constrained optimization problem.

Note that these $n+m$ equations are not necessarily easy to solve, for they may not be linear. For example, consider the problem

Minimize $\qquad (x_1-1)^2+(x_2-2)^2+(x_3-3)^2$

subject to $\qquad 2x_1+x_2^2=6$

$2x_1+x_2=9$

The Lagrangian function is given by

$$L(x_1,x_2,x_3,\lambda_1,\lambda_2)$$

$$=(x_1-1)^2+(x_2-2)^2+(x_3-3)^2+\lambda_1(2x_1+x_2^2-6)+\lambda_2(2x_1+x_3-9)$$

and the five necessary conditions are

$$\partial L/\partial x_1=2(x_1-1)+2\lambda_1+2\lambda_2=0$$

$$\partial L/\partial x_2=2(x_2-2)+2\lambda_1 x_2=0$$

$$\partial L/\partial x_3=2(x_3-3)+\lambda_2=0$$

$$\partial L/\partial\lambda_1=2x_1+x_2^2-6=0$$

$$\partial L/\partial\lambda_2=2x_1+x_3-9=0$$

These five nonlinear equations in five unknowns are rather difficult to solve. We note that the Lagrangian yields stationary points that are not necessarily maxima or minima.

Problems with inequality constraints. Consider the case of inequality constraints. Suppose that the problem is

Minimize $(x_1 - 3)^2 + (x_2 - 4)^2$

subject to $2x_1 + x_2 \leq 6$

Again create the Lagrangian with s the slack variable

$$L(x_1, x_2, \lambda) = (x_1 - 3)^2 + (x_2 - 4)^2 + \lambda(2x_1 + x_2 + s - 6)$$

The partial derivatives are

$$\partial L / \partial x_1 = 2(x_1 - 3) + 2\lambda$$
$$\partial L / \partial x_2 = 2(x_2 - 4) + \lambda$$
$$\partial L / \partial \lambda = 2x_1 + x_2 + s - 6$$

Now if $s > 0$, then L is minimized at $\lambda = -\infty$. Thus if λ is restricted to be at least 0 for the inequality constraint, then L is minimized at $\lambda = 0$. If $s = 0$, then it does not matter what λ is, since $L(x_1, x_2, \lambda) = f(x_1, x_2)$. Hence if (x_1, x_2, λ) minimizes L and $\lambda > 0$, then (x_1, x_2) minimized f and x_1, x_2 is feasible. Hence $\lambda(2x_1 + x_2 - 6) = 0$, since $\lambda = 0$ or $2x_1 + x_2 - 6 = 0$. Thus the conditions are

$$\frac{\partial}{\partial x_j} L(x_1, x_2) = 0$$

$$\lambda(2x_1 + x_2 - 6) = 0$$

Again we have $m + n$ nonlinear equations and $(m + n)$ unknowns. In this case λ is restricted in sign. The equations are

$$2(x_1 - 3) + 2\lambda = 0$$
$$2(x_2 - 4) + \lambda = 0$$
$$\lambda(2x_1 + x_2 - 6) = 0$$

with solution $x_1 = 7/5$, $x_2 = 16/5$, and $\lambda = 8/5$.

9.5. Summary

Nonlinear programming encompasses a wide variety of solution techniques. The easiest techniques to use are the search techniques, and it is seen that these methods behave in a fashion similar to the simplex method in the sense that the solution is improved at each interation. The final solution may be a *local optimum* or a *global optimum*, depending on the shape of the objective function.

The classical optimization techniques are some of the oldest techniques, but a major disadvantage is that they require the use of derivatives, which is a difficult task for computers. Their main usefulness is theoretical in the sense of the *Lagrange* and *Kuhn–Tucker* optimality conditions.

References and Selected Readings

Bazarra, M. S., and C. M. Shetty 1979. *Nonlinear Programming Theory and Algorithms.* New York: John Wiley and Sons, Inc.

Cooper, L., and D. Steinberg 1970. *Introduction to Methods of Optimization.* Philadelphia: Saunders.

Gardner, M. 1979. *Mathematical Circus.* New York: Knopf.

Luenberger, D. 1973. *Introduction to Linear and Nonlinear Programming.* Reading, Massachusetts: Addison-Wesley.

Wismer, D. A., and Chattergy. 1978. *Introduction to Nonlinear Optimization.* Amsterdam: North Holland.

Zangwill, W. 1969. *Nonlinear Programming: A Unified Approach.* Englewood Cliffs, New Jersey: Prentice-Hall, Inc.

Problems

1. Minimize the function $f(x)=x^2-5x+3$, $0\leqslant x\leqslant 10$ using

    (a) brute force
    (b) the bisection method
    (c) a Fibonacci search

    and 11 points for each search.

2. Minimize the function $g(y)=e^{-y}+2y^2$, $0\leqslant y\leqslant 1.2$ using

    (a) brute force
    (b) the bisection method
    (c) a Fibonacci search

    and 13 points for each search.

3. Find the minimum of the function $h(z)=z^4-2z^2$, $0\leqslant z\leqslant 8$ using

    (a) brute force
    (b) the bisection method
    (c) a Fibonacci search

    and 9 points for each search.

4. Use the method of cyclical coordinates and a starting point $(x,y)=(0,0)$ to minimize

$$f(x,y)=\left(x^3-y\right)^2+3(x-y)$$

5. Use the method of cyclical coordinates and a starting point $(x, y)=(1,1)$ to maximize

$$f(x, y)=x+2y+6xy-2x^2+2y^2$$

6. Use a grid search with 25 points for Problem 4 on the range $-2 \leqslant x, y \leqslant 2$.

7. Use a grid search with 36 points for Problem 5 on the range $-3 \leqslant x, y \leqslant 2$.

8. Solve Problem 1 using classical methods.

9. Attempt to solve Problem 2 using classical methods.

10. Solve Problem 3 using classical methods.

11. Solve Problem 4 using classical methods.

12. Solve Problem 5 using classical methods.

13. (a) Write the Lagrangian function for the following problem:

    Minimize $6x+8y$

    subject to $8x^2+3y=7$

    (b) Solve problem 13a using the Lagrange multipliers.

14. Prove that if $f(x)$ and $g(x)$ are convex then $h(x)=f(x)+g(x)$ is convex.

15. Show that $x^2-6x+y^2-8y+25$ is a convex function. [*Hint:* See problem 14.]

16. Show that $x=3, y=4$ is a stationary point for the function $x^2-6x+y^2-8y+25$.

17. Prove that if $n \geqslant 7$, then the bisection method is better than brute force for one dimensional optimization.

18. For the bisection method, the value $L/2^{(n-3)/2}$ is an upper bound for the final interval of uncertainty. Find a lower bound.
    [*Hint:* Consider the bisection method for the function $f(x)=1/x$.]

19. How is the constraint $|x| \leqslant 1$ incorporated into a linear program? [*Hint:* Let $x=x^+-x^-$ where $x^+, x^- \geqslant 0$.

Ten

Additional Topics in Mathematical Programming

10.1. Introduction

The previous chapters have all covered topics with only one decision maker and one objective. Now we examine situations where a decision made by one company affects a competing company. We also modify the interpretation with respect to some constraints, so that instead of saying, "The budget is strictly limited to $100,000," to allow for some flexibility we substitute the constraint, "The goal is to work within the budget of $100,000, if possible." The tools that are appropriate for these types of problems are game theory and goal programming, the subjects of the next two sections. The last section contains a solution method for assorted types of problems. The method is termed a greedy algorithm, since when this procedure is used decisions are made on a "next-best" basis rather than as a result of examining the overall picture. The greedy algorithm is simple to use and typically finds a good, if not an optimal, solution.

10.2. Game Theory

Up to this point an underlying assumption has been that the problems faced by a decision maker are all internal to his company. We now consider situations where two decision makers are in direct conflict with one another. Each decision maker must make one decision, and the profit for each decision maker is the result of the decisions made by both. Such conflicts are termed *two-person games*, and a special situation, called a *zero sum game*, prevails when one person's gain is exactly equal to the second person's loss. Consider the following example.

Deltair Airlines and Westmont Airlines are the only two airlines with direct flights from Philadelphia to Santa Fe. Currently, each airline has a

50% share of the Philadelphia/Santa Fe market. Now that the Federal Aviation Administration has reduced its regulation, each of the two airlines is considering the implementation of a new policy in order to increase its share of the market. Deltair is considering a 20% drop in prices, while Westmont is considering a 10% price reduction and a 25% increase in advertising. In Table 10.1 are the anticipated results of the implementation of the strategies. The results in the table indicate the net change in market share for any pair of strategies implemented. The change is from the perspective of Deltair, which means that Westmont's change is the negative of the number in the table. For example, if neither company implements a new strategy, then there is no change in the market share. If Deltair makes no change while Westmont advertises and reduces prices, Westmont increases its share of the market by 6% (and Deltair's decreases by the corresponding 6%). Alternatively, if Deltair decreases its prices while Westmont stands pat, Deltair increases its share of the market by 8%. Finally, if both companies implement a new strategy, Deltair gains 3% of the market.

Any two-person (or two-player) zero sum game can be represented by a table similar to Table 10.1. The general table contains m rows, one for each of player 1's m options, and n columns, one for each of player 2's n options. The number in row i, column j, represents the reward to player 1 if he takes action i and player 2 takes action j, while the reward to player 2 is the negative of this number. Table 10.2 is an example of a payoff table. For convenience, we look at the problem from player 1's point of view. The final solution, however, always dictates the optimal strategies for both players. Let us see how this works in the airline example.

Note that regardless of what strategy Westmont chooses, Deltair always prefers strategy 2. That is, Deltair prefers the 8% gain to no change and also prefers the 3% gain to a 6% loss. Similarly, Westmont prefers a 6% gain to no change and also prefers a 3% loss to an 8% loss. Therefore, the answer to this problem is that both companies must implement their new strategies D_2, W_2 and the outcome is a 3% gain for Deltair Airlines.

It is useful to consider the same example but with all of the numbers in the original table changed by 10%. That is, for all intents and purposes

Table 10.1. Airline Game

Deltair Strategies	Westmont Strategies[a]	
	W_1	W_2
D_1 (no change)	0	-6%
D_2 (20% price reduction)	8%	3%

[a]Strategy W_1 represents no change; strategy W_2 represents a 10% price reduction and a 25% increase in advertising.

Table 10.2. Return for Player 1 in the General Two-player Game

Player 1	Player 2				
	1	2	3	$\cdots$	n
1	0_{11}	0_{12}	0_{13}	$\cdots$	0_{1n}
2	0_{21}	0_{22}	0_{23}	$\cdots$	0_{2n}
3	0_{31}	0_{32}	0_{33}	$\cdots$	0_{3n}
$\vdots$	$\vdots$	$\vdots$	$\vdots$		
m	0_{m1}	0_{m2}	0_{m3}	$\cdots$	0_{mn}

Deltair gains an extra 10% of the market before any decisions are made. Notice from Table 10.3 that the answer is still the same. That is, the optimal decisions remain D_2 and W_2. In fact, a constant can always be added to *every* element of the payoff matrix without changing the problem. (It does change the value of the problem by an amount equal to the added constant but does not change the optimal strategies.) The importance of this fact is that adding a large enough constant always guarantees that the payoff matrix is greater than or equal to zero.

Let us consider another market share example with players 1 and 2, each having two strategies. The payoffs (to player 1) are given in Table 10.4. The answer to this problem is not as clear as the ones we have considered thus far. Player 1 does not prefer strategy A to strategy B all of the time, nor does he prefer strategy B to strategy A all of the time. At this point, it is not clear which strategy player 1 should choose. Consider player 2, who prefers strategy a to strategy b all of the time, since $60 < 65$ and $40 < 70$. Therefore, player 2 must choose strategy a. Furthermore, player 1 can reason this out, and by deducing that player 2 might prefer strategy a, player 1 realizes that he must choose strategy B. Thus the outcome is (B, a), yielding a value of 60 for player 1. Hence, all of the remaining work must include the fact that each player knows the reasoning of the other players. Also, if the optimal strategy of one player is clear, then the optimal strategy for the second player is also clear. Finally, let us consider the case in which it is not clear what either player should do.

Consider the game shown in Table 10.5. Each player has three strategies, and none of the strategies can be singled out as superior to any other strategy. Suppose that player 1 knows in advance what player 2 plans on

Table 10.3. Modified Airline Game

Deltair	Westmont	
	W_1	W_2
D_1	10%	4%
D_2	18%	13%

Table 10.4. Market Share Game

Player 1	Player 2	
	a	b
A	40	70
B	60	65

choosing. Then player 1 can choose his strategy. In this example, the strategies palyer 1 would choose would be C if player 2 chooses a, C if player 2 chooses b, and B if player 2 chooses c. The values of (C, a), (C, b), and (B, c) are 7, 6, and 9, respectively. If the rules of the game were that player 2 must choose first, then he would have to choose strategy b, since $6 < 7 < 9$. This new game with player 2 choosing first is unfair to player 2. It stands to reason that for a fair game, where both players select at the same time, player 2 should do no worse than he does in the unfair game. Therefore, the value of this game to player 2 can be no worse than -6. Alternatively, the value of this game to player 1 is no better than $+6$.

Since the zero sum game is symmetrical in form, let us consider the case in which player 1 chooses first and player 2 knows the choice. Player 2 chooses b if player 1 chooses A, a if player 1 chooses B, and c if player 1 chooses C. The values of (A, b), (B, a), and (C, c) are 5, 4, and 3, respectively. Knowing this, player 1 chooses A, guaranteeing a value to him of 5. Thus player 1 can guarantee a value of at least 5 but would prefer 6. Player 2 can guarantee losing no more than 6, but wants to lose only 5.

Note that if either player has a predetermined strategy, the other player can take advantage of this fact. Therefore, it is essential that each player not let the other know his strategy in advance by not knowing it himself! Each player must choose his strategy randomly. Suppose that player 1 selects strategy A, B, or C with probabilities p_A, p_B, and p_C, respectively. Of course $p_A + p_B + p_C = 1$. The game value then if player 2 chooses strategy a is $6p_A + 4p_B + 7p_C$. If player 2 chooses strategy b, the game value is $5p_A + 5p_B + 6p_C$, while if player 2 chooses strategy c, the game value is $7p_A + 9p_B + 3p_C$. What player 1 would like to do is to maximize all three of these game values at the same time. Thus the problem for player 1 is

$$\text{Maximize } 6p_A + 4p_B + 7p_C$$

Table 10.5 Three Strategy Game

Player 1	Player 2		
	a	b	c
A	6	5	7
B	4	5	9
C	7	6	3

and

$$\text{Maximize } 5p_A + 5p_B + 6p_C$$

and

Maximize $\qquad\qquad 7p_A + 9p_B + 3p_C$

subject to $\qquad\qquad p_A + p_B + p_C = 1$

$$0 \leqslant p_j \leqslant 1 \qquad j = A, B, C$$

This would be a linear programming problem except for the fact that there are three objective functions. Furthermore, for any p_A, p_B, and p_C, player 2 wishes to minimize the payoff and chooses the smallest of the three values. For example, if for some reason player 1 decides that $p_A = p_B = p_C = \frac{1}{3}$, then

$$6p_A + 4p_B + 7p_C = 17/3$$
$$5p_A + 5p_B + 6p_C = 16/3$$

and

$$7p_A + 9p_B + 3p_C = 19/3$$

which means that player 2 chooses strategy b and player 1 gains 16/3. This is better than the 5 that has already been guaranteed to player 1 but the question remains whether or not player 1 can do even better. Since player 2 always chooses the strategy that yields the minimum of the three values, player 1's objective should be to maximize this minimum. That is, player 1's real problem is

Maximize $\{\text{minimum}\{6p_A + 4p_B + 7p_C, 5p_A + 5p_B + 6p_C, 7p_A + 9p_B + 3p_C\}\}$

subject to $\qquad\qquad p_A + p_B + p_C = 1$

$$0 \leqslant p_j \leqslant 1 \qquad j = A, B, C$$

Let us express the max-min objective function by a simple transformation. Let v be the value of the game. Then player 1 is interested in maximizing v while making certain that each of the three values in the objective function is at least equal to v. Thus the problem is

Maximize $\qquad\qquad\qquad v$

subject to $\qquad\qquad 6p_A + 4p_B + 7p_C \geqslant v$

$$5p_A + 5p_B + 6p_C \geqslant v$$

$$7p_A + 9p_B + 3p_C \geqslant v$$

$$p_A + p_B + p_C = 1$$

$$p_A \qquad\quad \leqslant 1$$

$$p_B \quad\ \leqslant 1$$

$$p_C \leqslant 1$$

$$p_A, p_B, p_C \geqslant 0$$

Notice that the value of the objective function appears as a variable. Similarly, if p_a, p_b, and p_c are the percentages of time that player 2 uses strategies a, b, and c, player 2 is interested in the problem

$$
\begin{aligned}
\text{Minimize} \qquad & v \\
\text{subject to} \qquad & 6p_a + 5p_b + 7p_c \leqslant v \\
& 4p_a + 5p_b + 9p_c \leqslant v \\
& 7p_a + 6p_b + 3p_c \leqslant v \\
& p_a + p_b + p_c = 1 \\
& p_a \quad \leqslant 1 \\
& \quad p_b \leqslant 1 \\
& \quad p_c \leqslant 1 \\
& p_a, p_b, p_c \leqslant 0
\end{aligned}
$$

Notice that for either problem, since p_a, p_b, and p_c must add to 1 and each must be greater than zero, the constraints $p_a \leqslant 1$, $p_b \leqslant 1$, $p_c \leqslant 1$ are redundant. It follows that any two-player zero sum game can be represented and solved as a linear programming problem.

Let us rewrite the foregoing linear programming problem by dropping the three redundant constraints, changing minimize v to maximize $-v$, and multiplying the equality constraint by -1. Hence player 2's problem can be expressed as

$$
\begin{aligned}
\text{Maximize} \qquad & -v \\
\text{subject to} \qquad & 6p_a + 5p_b + 7p_c - v \leqslant 0 \\
& 4p_a + 5p_b + 9p_c - v \leqslant 0 \\
& 7p_a + 6p_b + 3p_c - v \leqslant 0 \\
& -p_a - p_b - p_c = -1 \\
& p_a, p_b, p_c \geqslant 0
\end{aligned}
$$

Now, using the rules of Section 4.4 we can create the dual of this linear program. Since there are four constraints, there must be four dual variables, which we label y_1, y_2, y_3, and w. The dual problem is

$$
\begin{aligned}
\text{Minimize} \qquad & -w \\
\text{subject to} \quad & 6y_1 + 4y_2 + 7y_3 - w \geqslant 0 \\
& 5y_1 + 5y_2 + 6y_3 - w \geqslant 0 \\
& 7y_1 + 9y_2 + 3y_3 - w \geqslant 0 \\
& -y_1 - y_2 - y_3 = -1 \\
& y_1, y_2, y_3, w \geqslant 0
\end{aligned}
$$

Note that this is precisely player 1's problem.

In summary, either player can represent his problem as a linear program and solve the problem using the standard techniques. Furthermore, the dual to each player's problem is exactly the linear program that his opponent must solve!

10.3. Goal Programming

In the traditional linear programming model there is only one objective, and all other considerations are implemented through constraints. Restricting models to having only one objective may be very limiting. For example, consider the dilemma confronting a political candidate seeking election to whom it is not immediately clear whether he should maximize the number of votes received subject to a budget constraint, or minimize the amount of money spent subject to the constraint of getting at least 51% of the votes. In reality, the average politician does not have a budget constraint, since if an extra $1000 would get an additional, say 10% of the crucial votes, the candidate would "somehow manage" to get the $1000. The candidate actually has a *set* of goals, including the potential votes and the money spent. Let us address the general issue of how to approach a problem that has more than one objective.

Consider again the Exclusive Furniture Company problem as given in Chapter 4. The problem is

$$\text{Maximize} \qquad\qquad 5x+7y$$

$$\begin{array}{lll} \text{subject to} & 2x+5y \leqslant 51 & \text{(materials constraint)} \\ & 3x+2y \leqslant 42 & \text{(man-hour constraint)} \\ & x+y \geqslant 14 & \text{(total number of pieces)} \\ & y \geqslant 7 & \text{(at least 7 tables)} \\ & x, y \geqslant 0 \end{array}$$

In Chapter 4 the formulation of the Exclusive Furniture Company problem is such that the company *must* make at least seven tables. Suppose now that instead of viewing the number of tables as a constraint, we regard the seven tables as a goal. That is, the Exclusive Furniture Company would like to maximize profit and would like to make seven tables subject to constraints on materials, man-hours, and total number of pieces. Furthermore, since the reference is to goals, let us be specific and state that the profit goal is $100. That is, the two goals are

$$\text{Goal 1} \qquad\qquad 5x+7y=100$$

and

$$\text{Goal 2} \qquad\qquad y=7$$

The immediate question is how to incorporate the two goals into a linear programming type format. The tools for converting the problem into the proper form are simply the slack and surplus variables used previously. As an example, consider the pair $x=10$, $y=4$. Note that $(10,4)$ satisfies the three constraints. With respect to the goals the values are

$$5x+7y=51+28=79 \text{ dollars}$$

and

$$y=4 \text{ tables}$$

Hence a complete solution must indicate the shortage with respect to both goals. Express Goal 1 as

$$5x+7y+s_1=100$$

where in this particular case s_1 would be 23, the amount below the goal of 100. Furthermore, in general it is possible to be above or below a goal. For this reason both a slack variable and a surplus variable are needed. Thus

$$\text{Goal 1} \qquad 5x+7y+s_1^- -s_1^+=100$$

where s_1^- represents the amount below the goal and s_1^+ represents the amount above the goal. Similarly, we have

$$\text{Goal 2} \qquad y+s_2^- -s_2^+=7$$

and s_2^- is the number of tables produced below seven and s_2^+ is the number of tables produced above seven. Note that all of the new variables that have been added are restricted in sign to being nonnegative.

At this point the two new equations (Goal 1 and Goal 2) are used to define the new variables. Hence they enter the problem as constraints. The problem that remains is to define an objective function. There exists an elegant solution to this problem. First, assume that having a surplus profit or a surplus number of tables does not concern the company. Suppose that the penalty for every table below the goal is $10. Obviously, the penalty for every dollar below the goal of $100 is $1. Therefore the ratio of penalties between tables and budget is 10 to 1. The objective function must represent this relationship, which is that the cost of being short in tables is ten times the cost of being short in profit. Hence the complete problem is

Minimize $\qquad\qquad\qquad\qquad s_1^- +10s_2^-$

subject to $\quad 5x+7y+s_1^- -s_1^+ = 100 \qquad$ (Goal 1)

$\qquad\qquad\qquad y+s_2^- -s_2^+ = 7 \qquad\qquad$ (Goal 2)

$\qquad\qquad\qquad\quad 2x+5y \leqslant 51 \qquad\qquad$ (materials)

$\qquad\qquad\qquad\quad 3x+2y \leqslant 42 \qquad\qquad$ (man-hours)

$\qquad\qquad\qquad\qquad\quad x+y \geqslant 14 \qquad\qquad$ (total number of pieces)

$$x, y, s_1^-, s_1^+, s_2^-, s_2^+ \geqslant 0$$

This is a standard linear programming problem! (See Problem 10.)

The formulation just completed requires the establishment of weights for each of the competing goals, that is, the weight associated with s_1^+ is zero, with s_2^+ is zero, with s_1^- is one and with s_2^- is ten. There is an alternative approach that is worth considering. For convenience consider a smaller problem eliminating the restrictions on man-hours and total number of pieces.

$$\text{Minimize} \qquad\qquad f(s_1^-, s_2^-)$$

$$\text{subject to} \qquad\qquad 5x + 7y + s_1^- - s_1^+ = 100$$

$$y + s_2^- - s_2^+ = 7$$

$$2x + 5y \leqslant 51$$

$$x, y, s_1^-, s_1^+, s_2^-, s_2^+ \geqslant 0$$

Notice that a general objective function of the two important variables is used. An alternative approach that is more general is (instead of weighting the importance of s_1^- and s_2^-) to express the objective as:

First satisfy Goal 1 and if this is accomplished, then satisfy Goal 2.

For convenience write

$$\text{Minimize } f(s_1^-, s_2^-) = p_1 s_1^- + p_2 s_2^- \quad \text{or} \quad \text{maximize } -p_1 s_1^- - p_2 s_2^-$$

However, p_1 is not a multiplier but means that p_1 has first priority and p_2 has second priority. Alternatively, it is convenient at times to think of this relationship as $p_2/p_1 \to 0$ (close to zero). Now modify the simplex procedure so that it satisfies the goals according to their priorities. Table 10.6 contains the problem in the form of a tableau. Let s_1^- and s_2^- be the basic variables in rows 1 and 2. (This is easier than adding artificial variables.) However, since s_1^- and s_2^- appear in the last row, perform row operations on the final row so that the columns s_1^- and s_2^- have zeros in every row except the row in which they are basic. This leads to Table 10.7. The

Table 10.6. Goal Programming: First Tableau

Row	Basic	Value	x	y	s_1^-	s_1^+	s_2^-	s_2^+	s_3
1	s_1^-	100	5	7	1	-1	0	0	0
2	s_2^-	7	0	1	0	0	1	-1	0
3	s_3	51	2	5	0	0	0	0	1
			0	0	p_1	0	p_2	0	0

Table 10.7. Goal Programming: Second Tableau

Row	Basic	Value	x	y	s_1^-	s_1^+	s_2^-	s_2^+	s_3
1	s_1^-	100	5	7	1	-1	0	0	0
2	s_2^-	7	0	1	0	0	1	-1	0
3	s_3	51	2	5	0	0	0	0	1
		$-100p_1-7p_2$	$-5p_1$	$-7p_1-p_2$	0	p_1	0	p_2	0

coefficients in the last row for the nonbasic variables are

Variable	Coefficient
x	$-5p_1$
y	$-7p_1-p_2$
s_1^+	p_1
s_2^+	p_2

This means that if x enters the basis, Goal 1 improves by $5p_1$; if y enters the basis, Goal 1 improves by $7p_1$ and Goal 2 improves by p_2. Similarly, if s_1^+ or s_2^+ enters, Goal 1 and 2 worsen, respectively, by p_1 and p_2. The immediate concern is improving goal 1, since it has the highest priority. Let x enter the basis. The new tableau is given in Table 10.8. From this tableau it can be seen that the first goal is met. Also, y is the only variable that can improve Goal 2. Therefore let y enter the basis which yields Table 10.9. From this table it is observed that only s_1^- can improve Goal 2. However if s_1^- enters the basis Goal 1 will fail. Therefore this is the final solution.

Table 10.8. Goal Programming: Third Tableau

Row	Basic	Value	x	y	s_1^-	s_1^+	s_2^-	s_2^+	s_3
1	x	20	1	$\frac{7}{5}$	$\frac{1}{5}$	$-\frac{1}{5}$	0	0	0
2	s_2^-	7	0	1	0	0	1	-1	0
3	s_3	11	0	$\frac{11}{5}$	$-\frac{2}{5}$	$\frac{2}{5}$	0	0	1
		$-7p_2$	0	$-p_2$	p_1	0	0	p_2	0

Table 10.9. Goal Programming: Final Tableau

Row	Basic	Value	x	y	s_1^-	s_1^+	s_2^-	s_2^+	s_3
1	x	13	1	0	$\frac{5}{11}$	$-\frac{5}{11}$	0	0	$\frac{7}{11}$
2	s_2^-	2	0	0	$\frac{2}{11}$	$-\frac{2}{11}$	1	-1	$-\frac{5}{11}$
3	y	5	0	1	$-\frac{2}{11}$	$\frac{2}{11}$	0	0	$\frac{5}{11}$
		$2p_2$	0	0	$p_1-\frac{2}{11}p_2$	$\frac{2}{11}p_2$	0	p_2	$\frac{5}{11}p_2$

10.4. Greedy Algorithms

Up to this section, all of the algorithms that have been presented are guaranteed to find the optimal solution if one exists. Often a problem is so complex that finding the optimal solution might be too expensive. In these cases, we might want to settle for a technique that leads to a good, but not necessarily optimal, solution, provided that the technique is relatively efficient and economical. These approaches are termed heuristic methods. A disadvantage of heuristic methods is that it is not generally known whether the solution obtained by the heuristic is optimal. One of these heuristic methods, called a *greedy algorithm*, is presented in this section.

The greedy algorithm is possibly the most reasonable general solution method that is available for assorted types of problems. Assume that a problem requires n decisions that are made sequentially. The immediate reward for the jth decision is r_j. The reward for all of the decisions $j, j+1,\ldots,n$ is P_j. Thus, as in dynamic programming, we decompose the problem into the general form

$$P_j = r_j + P_{j+1}$$

where P_j refers to the total from decision j until the end and r_j refers to the immediate reward for making a decision. However, whereas when using dynamic programming the consequences of r_j and the new state are examined, we now consider only the reward r and make the decision based solely on the immediate return. Such an algorithm is termed greedy because it is a short-sighted algorithm that maximizes immediate rewards regardless of future consequences. Before discussing the virtues and pitfalls of greedy algorithms, let us consider an example.

Figure 10.1 is the network presented earlier in Chapter 8. Again suppose that we are interested in finding the shortest path from point A to point G. The greedy algorithm prescribes beginning at A and always advancing to the closest adjacent unvisited node. The closest point to A is point B, since $2 < 3 < 5$. From B move to C, since $2 < 8$. From C travel to D, since $1 < 2$; then proceed to E, F, and finally G. The greedy algorithm has generated the path A–B–C–D–E–F–G. Although it is true that in this case the path is suboptimal, an important consideration is how easy it is to apply the algorithm. The greedy algorithm always generates an upper bound on the cost and, surprisingly, the upper bound many times is close to or even equal to the optimal solution. The computations required for the greedy algorithm are substantially fewer than those needed for the algorithms examined in prior chapters. Furthermore, there are some problems for which the greedy method actually computes the optimal solution. The next two subsections present two of these problems, and the last subsection presents a greedy algorithm for the traveling salesman problem.

Before discussing these topics, it should be noted that Vogel's approximation method of Section 5.2 is a heuristic algorithm. Shipping allocations are

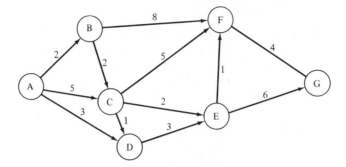

Figure 10.1. Shortest Path Problem

made on a next-best (greedy) basis. With Vogel's approximation method, the optimal solution is sometimes found, but if not, the starting solution is a good one.

10.4.1. The Minimum Spanning Tree Problem

The planned locations of computer terminals that are to be installed in a multistory building are given in Figure 10.2. Terminal A is the computer itself and phone cables must be wired along some of the indicated branches in order that there be a connected path from every terminal back to A. The numbers along the arcs represent the costs (in hundreds of dollars) of installing the lines between terminals. Since operating costs are very low, the company would like to find the branches that should be installed in order to minimize total installation costs.

Figure 10.3 presents a feasible, though not necessarily optimal, solution that costs $2000. Since any feasible solution looks somewhat like the tree of

Figure 10.2. Minimum Spanning Tree Problem

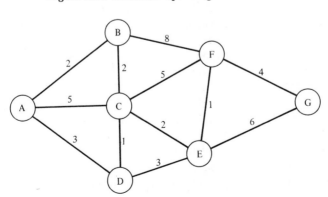

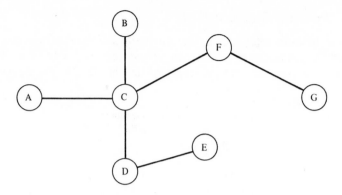

Figure 10.3. Feasible Solution to the Spanning Tree Problem

Figure 10.3, the problem is termed the minimum spanning tree problem. The easiest method of solution for this problem is a greedy method.

Begin at any terminal and find the branch to the nearest unconnected terminal. This branch is part of the tree. From these two terminals find the branch to the nearest unconnected terminal and add this branch to the tree. Continue in this fashion until all terminals are connected.

Let us begin arbitrarily at node C in the example in Figure 10.2. The nearest terminal to C is D; CD is a branch on the tree. The nearest terminals to both C and D are B and E; we arbitrarily chose E. Next, find F closest to C, D, and E, and finally choose B, then A and G. The spanning tree is as given in Figure 10.4 and the total cost is $1200. This happens to be the optimal solution even though the greedy method has been used to solve the problem. The greedy method always finds the minimum spanning tree.

Figure 10.4. Optimal Spanning Tree

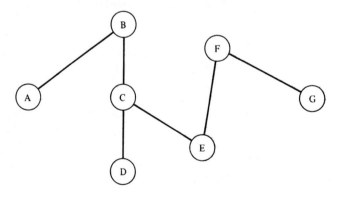

10.4.2. Job Shop Sequencing

Consider an auto repair shop that currently has five cars waiting to be painted. Only one car at a time can be painted. The accompanying table designates the five cars and the length of time it takes to paint each one.

Car	Painting time (hours)
A	6
B	2
C	5
D	4
E	3

The objective is to find the order in which the cars should be painted so that the average time it takes to get a car out of the shop is minimized. In other words, if T_i is the completion time of job i then the objective is to minimize $(1/5)\Sigma_{i=1}^5 T_i$. For example, one schedule is to paint the cars in the order A, B, C, D, E. For this sequence the completion times are as given below.

Car	Completion time (hours)
A	6
B	8
C	13
D	17
E	20
	64

The total completion time is 64; hence, the average completion time per car is 64/5. Obviously, minimizing total or average completion times yields the same schedule. Notice that there are $5! = 120$ possible sequences. In fact, for the general case of n jobs there are $n!$ sequences from which to choose.

Consider the general problem where t_j represents the processing time for job j. Let us consider any two jobs i and j and the two sequences ij and ji, where ij indicates that job i immediately precedes job j and j_i indicates that job j immediately precedes job i. Let $F(s)$ represent the total completion time of the jobs in the sequence s. Then for adjacent i, j with starting time T_0, $F(ij) = T_0 + t_i + (T_0 + t_i + t_j)$ and $F(ji) = T_0 + t_j + T_0 + t_j + t_i$. Hence, $F(ij) - F(ji) = t_i - t_j$. If $t_i - t_j < 0$, then the schedule ij is better than the schedule ji. In other words, if $t_i < t_j$, job i should precede job j. Since this is true for any two adjacent jobs, the optimal schedule is to start with the shortest job, then proceed to the second shortest job, and continue in this way; the longest job is scheduled last. In other words, be greedy!

The optimal schedule is found very simply to be B, E, D, C, A. The completion times are shown in the accompanying table.

Car	Completion time (hours)
B	2
E	5
D	9
C	14
A	20
	50

Notice that the greedy schedule finishes the cars $2.8 = 64/5 - 50/5$ hours earlier (on the average) than the arbitrary schedule A, B, C, D, E, and finishes the cars 4 hours earlier (on the average) than the worst schedule, A, C, D, E, B, which takes $70/5$.

The concept can be extended further. Suppose that associated with each job is a weight, w_j, which represents the relative importance of job j, and that the objective is to minimize the sum of the weighted completion times. That is, if T_j is the time at which job j is completed, the objective is to minimize

$$\sum_{j=1}^{n} w_j T_j$$

For any two adjacent jobs i, j with starting time T_0, the weighted completion times $F(ij) = w_i(T_0 + t_i) + w_j(T_0 + t_i + t_j)$ and $F(ji) = w_j(T_0 + t_j) + w_i(T_0 + t_j + t_i)$. Thus $F(ij) - F(ji) = w_j t_i - w_i t_j$, so if $w_j t_i < w_i t_j$, schedule job i before job j. Another way to state this is, if $t_i/w_i < t_j/w_j$, schedule job i before job j. Again the answer is to be greedy, but instead of using t_j as the criterion, use t_j/w_j (time per weight). For example, given the following problem:

Car	Time	Weight
A	6	2
B	2	1
C	5	4
D	4	1
E	3	2

compute the (inversely) weighted times (t_j/w_j):

Car	t_j/w_j
A	3
B	2
C	1.25
D	4
E	1.5

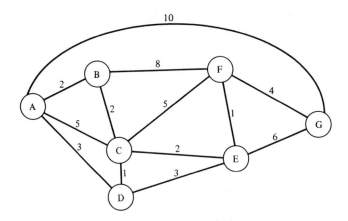

Figure 10.5. The Traveling Salesman Problem

and find that the optimal schedule is C, E, B, A, D, with a weighted cost of 98. This is considerably less than the weighted cost of 146 generated by the worst schedule, D, A, B, E, C.

10.4.3. A Traveling Salesman Example

Figure 10.5 is a network representation of seven cities and the roads linking these cities. A salesman whose home office is in city A must visit each of the cities exactly once and then return home. He would like to do so in the order that minimizes the total distance traveled. The problem is a very difficult one to solve in general. It is possible to use a dynamic programming approach but this has proved unsuccessful for problems with more than 20 cities. Hence, a greedy algorithm is a method that yields a good, but not necessarily optimal, solution. Very simply, we begin at his home office A and always proceed to the nearest unvisited city. For this problem the solution is A–B–C–D–E–F–G–A, with a total distance of 23.

10.5 Summary

The scope of mathematical programming is extensive and it is impossible to present every idea in a single book. The major emphasis of the first nine chapters has been on the development of algorithms that find optimal solutions to problems. In this last chapter, however, it is shown that there exist very quick, efficient *greedy methods* based on simple common sense for many problems that find good, but not necessarily optimal, solutions.

In addition, it is seen that the interpretation of constraints can be changed so that goals are used along with the cost of not meeting these *goals*. This modification makes linear programming considerably more general.

Finally, the very beginnings of *game theory* are presented. The general decision-making problem may very well involve more than one company, in which case it is possible that the only useful analysis is one that includes more than one decision maker and an $n \times m$ payoff matrix.

References and Selected Readings

Conway, R. W., W. L. Maxwell, and L. W. Miller 1967. *Theory of Scheduling.* Reading, Massachusetts: Addison-Wesley.

Hesse, R., and G. Woolsey 1980. *Applied Management Science.* Chicago, Illinois: Science Research Associates.

Hillier, F., and G. Lieberman 1980. *Introduction to Operations Research*, 3d ed. San Francisco: Holden-Day.

Lee, S. M. 1972. *Goal Programming for Decision Analysis.* Philadelphia: Auerbach Pub. Inc.

Luce, R. D., and H. Raiffa 1957. *Games and Decisions.* New York: John Wiley and Sons, Inc.

Williams, J. D. 1966. *Compleat Strategyst.* New York: McGraw-Hill.

Woolsey, R. E. D., and H. S. Swanson 1975. *Operations Research for Immediate Application.* New York: Harper & Row.

Problems

1. Develop a linear programming model to solve the following two-player zero sum game.

	Player II	
Player I	a	b
A	4	7
B	6	3
C	5	4

2. Solve each of the following two-player zero sum games.

(a)

	Player II		
Player I	a	b	c
A	5	10	4
B	4	5	2
C	4	8	3
D	6	7	5

(b)

	Player II		
Player I	a	b	c
A	5	8	7
B	4	6	4
C	3	5	5

(c)

	Player II	
Player I	a	b
A	4	8
B	8	9
C	5	10
D	6	5

(d)

	Player II			
Player I	a	b	c	d
A	4	10	2	2
B	5	7	5	7
C	6	4	2	4
D	8	5	4	6

3. Penthome Incorporated is setting up the production of at least 5 pounds of a product that contains two different materials in a proportion of at most 7 pounds of material A per 8 pounds of material B. The cost is $.30 and $.60 per pound of material A and material B, respectively.

   Suppose Penthome has two goals: the first is to use all available stock of material A, which is actually 3 pounds; the second is to attain a production cost of $3.00.

   First, suppose that the trade-off between the goals can be resolved by specifying that the cost of using less than 3 units of A is 4 times the cost of having a production cost beyond $3.00. Second, suppose that all the company wants is to satisfy goal 1 first, and then goal 2. In both cases, determine the optimal mix of materials A and B.

4. Given the following three goals in order of priority, find the values of x_1 and x_2 that maximize the attainment of such goals:

   goal 1 $6x_1 + 4x_2 = 200$
   goal 2 $4x_1 - 2x_2 = 40$
   goal 3 $5x_1 + 2x_2 = 60$

5. Obtain a solution for the following shortest-route problem by using the greedy algorithm, and then compare it to the solution obtained by using dynamic programming.

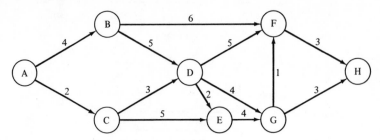

6. Apply the greedy algorithm to solve the following minimum spanning tree problem.

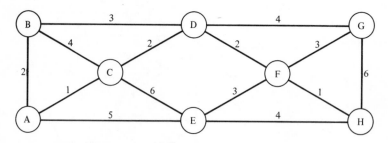

7. Given six tasks and their completion times (see the table below), find the schedule that minimizes the total completion time for all tasks.

Task	Time
1	15
2	20
3	18
4	9
5	10
6	13

8. Suppose that in Problem 7 the following weights are assigned to each of the tasks.

Task	Weights
1	4
2	2
3	5
4	4
5	3
6	4

find the optimal schedule, that minimizes average weighted completion time.

9. Find a solution to the following two traveling salesman problems by using the greedy algorithm.

(a)

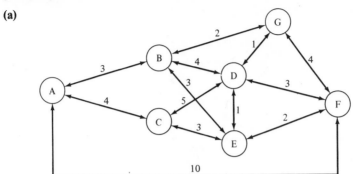

(b)

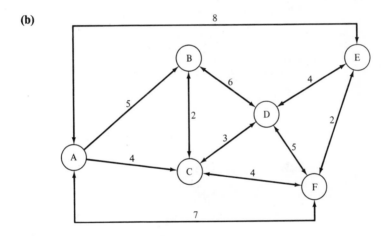

10. Solve the goal program in Section 10.3.

Answers
to Selected Problems

Chapter 2

9. $z = 1325$; $x_1 = 17.5$; $y = 7.5$

10. $z = 2.15$; $x = 21.5$; $y = 0$

11. $z = 1000$; $x = 8$; $y = 3$

12. $z = 32857.1$; $x = 1.42857$; $y = 2.14286$

13. $z = 46$; $x = 1$; $y = 7$

14. $z = 136.8$; $x = 2.2$; $y = 5.6$

15. $z = 17$; $x = 10$; $y = 7$

16. $z = 80.8$; $x = 3.6$; $y = 3.2$

17. $z = 72$; $x = 3$; $y = 6$

18. $z = 20$; $x = 0$; $y = 1$

19. $z = 10$; $x = 0$; $y = 2$

20. Solution unbounded

21. $z = 3.33$; $x = 0.167$; $y = 0.8333$

22a. $(0,3)$; $(0,1)$; $(0,0)$; $(0,-2)$; $(2,1)$; $(4,1)$; $(6,1)$; $(3,0)$; $(4,0)$; $(4,-1)$; $(10/3, -2/3)$

22b. $(0,1)$; $(2,1)$; $(3,0)$; $(0,0)$

22c.

x	y	15x	10y	15x + 10y
0	0	0	0	0
0	1	0	10	10
2	1	30	10	40
3	0	45	0	45

23. a. $\binom{n}{m}$

   b. Yes, three or more lines may intersect at the same point or two or more constraints may be parallel.

24. a. $z = 30$; $z = 7.5$; $z = 0$; $z = -30$;

   b. $z = 17/3$; $z = 50/3$; $z = 9$;

   c. $z = 5$; $z = 0$; $z = 5$

Chapter 3

1. $z = 45$; $x_1 = 0$; $x_2 = 2.5$; $x_3 = 1.875$

2. Solution unbounded

3. $z = 138.571$; $x_1 = 0$; $x_2 = 8.57143$; $x_3 = 34.2857$; $x_4 = 15.7143$

4. $z = 104$; $x_1 = 11$; $x_2 = 9$; $x_3 = 0$; $x_4 = 7$

5. $z = 100$; $x_1 = 50$; $x_2 = 50$; $x_3 = 0$

6. $z = 1200$; $x_1 = x_2 = x_3 = 0$; $x_4 = 200$

7. Solution unbounded

8. $z = 35$; $x_1 = 0$; $x_2 = 22\ 1/2$; $x_3 = 5$

9. $z = 18.6154$; $x_1 = 5.84615$; $x_2 = 0$; $x_3 = 1.38462$; $x_4 = x_5 = 0$.

10. $z = 216$

11. $z = 1380.95$; $x_1 = 0$; $x_2 = 66\ 2/3$

12. $z = 6440$; $x_1 = 920$; $x_2 = 0$

13. $z = 32857.1$; $x_1 = 1.42857$; $x_2 = 2.14286$
   a. if $a_{rs} > 0$
   b. if $a_{rs} \leqslant 0$

15. $x_r = 0$; $c_s < 0$

16. Finite optimal solution with an unbounded feasible region

17. The solution is feasible

18. $z = -5.9$; $x_1 = 1.3$; $x_2 = 0$; $x_3 = .4$

19. $z=15$; $x_1=0$; $x_2=8.33$; $x_3=6.67$; $x_4=x_5=0$

20. Solution unbounded

Chapter 4

1. $z=4$; $x_1=1$; $x_2=1$; $x_3=0$

1a. $z=2$; $x_1=1$; $x_2=x_3=0$

1b. $z=4$; $x_1=1$; $x_2=1$; $x_3=0$

2a. $z=24$; $x_1=2$; $x_2=0$; $x_3=1$

2b. $z=25$; $x_1=x_2=5$; $x_3=0$

2c. $z=152\ 1/2$; $x_1=x_2=0$; $x_3=20$; $x_4=17\ 1/2$

3a. $z=25$; $x_1=x_2=5$; $x_3=x_4=0$

3b. $z=25$; $x_1=x_2=5$; $x_3=x_4=0$

4a. $z=62\ 1/2$; $x_1=x_2=x_3=0$; $x_4=7\ 1/2$; $x_5=20$

4b. $z=152\ 1/2$; $x_1=x_2=0$; $x_3=20$; $x_4=17\ 1/2$; $x_5=0$

5a. $z=6$; $x_1=x_2=1$; $x_3=0$

5b. $z=6$; $x_1=2$; $x_2=1$; $x_3=0$

5c. $z=2$; $x_1=1$; $x_2=x_3=0$

8. $z=3$; $x_1=1.5$; $x_2=0$ (Primal)

$z=3$; $y_1=0$; $y_2=1$ (Dual)

Chapter 5

1. $z=15945$

2. $z=103620$

3. $z=6600$

4. $z=16400$

5. $z=83$

6. Distance$=57900$ miles; Min cost$=\$33,950$

7a. Max $z=180$

7b. $z=745$

7c. $z=415$

7d. $z=351$

7e. $z=355$

7f. $z=1650$

8. $u_1+v_1\leqslant 3$

$u_1+v_2\leqslant 4$

$u_1+v_3\leqslant 6$

$u_2+v_1\leqslant 2$

$u_2+v_2\leqslant 4$

$u_2+v_3\leqslant 5$

Max $400u_1+250u_2+250v_1+300v_2+150v_3$

u_i, v_j unrestricted

10, 11. Vogel's method is predicated on the assumption that if the lowest cost in a row or column is not used then the second lowest cost will be used. However, there is no guarantee that, in fact, the second lowest cost will be used. It may be the third, fourth, etc. cost that is actually used.

$x_{11}=200$, $x_{13}=50$, $x_{22}=150$, $x_{32}=250$, $x_{42}=100$, $x_{43}=100$, $x_{53}=50$, $x_{54}=200$, $x_{63}=150$

12. $z=93$

Chapter 6

1. $z=9$

2. $z=609$

3. $z=26$

4. $z=75$

5. $z=15$

6. $z=64$

7. $z=238$

8. $z=430$

9. $z=6$

10. $z=42$

11. Bottleneck$=4$;

12. Bottleneck$=15$;

Chapter 7

2. $z=4320$; $x_1=x_2=x_3=x_4=x_5=x_6=x_7=x_8=x_9=0$; $x_{10}=24$

3. Min $z=12$; $x_1=4$

4. $z=32$; $x_1=8$

5. $z=20$; $x_1=4$

6. $z=5$ (Multiple solutions)

x_1	x_2
3	2
5	0
4	1

9. $z=26$; $x_1=3$; $x_2=1$; $x_3=2$

10. $z=100$; $x_2=5$

Chapter 8

1a. Shortest path $=A-1-5-7-B$

1b. Shortest path $=A-1-3-5-7-B$

2. (2 Alternatives)

Region	Adv.	Eff.	Region	Adv.	Eff.
4	5	22	4	4	18
3	4	20	3	4	20
2	0	0	2	1	4
1	1	6	1	1	6
		$\overline{48}$			$\overline{48}$

3. (2 Alternatives)

	Projet	$Invest.	$Revenue	Proj.	$Inv.	$Rev.
26	3	0	0	3	6	16
	2	5	12	2	4	10
	1	5	14	1	0	0
			$\overline{26}$			$\overline{26}$

4. $41,000   1500 @ $16,000
            2500 @ $25,000
            $\overline{\$41,000}$

5.

Machinery	Tons	Value
3	0	0
2	0	0
1	10	1
—	—	$\overline{1}$

6. $x_{15}=1$; all other $x_{ij}=0$

7. $x_1=0, x_2=3, x_3=0$

8. $x_1=0, x_2=5, x_3=0$

9. $x_1=10, x_2=0, x_3=0$

10. $x=1, y=7$

Chapter 9

1a. $2\leqslant x^*\leqslant 3$

1b. $2.1875\leqslant x^*\leqslant 2.8125$

1c. $360/144\leqslant x^*\leqslant 370/144$ or $350/144\leqslant x^*\leqslant 360/144$

2a. $0.1\leqslant x^*\leqslant 0.3$

2b. $0.1875\leqslant x^*\leqslant 0.2250$

2c. $0.2037135\leqslant x^*\leqslant 0.2068966$

3a. $0\leqslant z^*\leqslant 2$

3b. $0.5\leqslant z^*\leqslant 1.5$

3c. $48/55\leqslant z^*=56/55$

4. The solution approaches $x^*=\sqrt{0.5}$, $y^*=1.5-0.5\sqrt{0.5}$ as the number of iterations increases

5. $x^*\to -4/26, y+\to -7/26$

6. $-1\leqslant x^*\leqslant 1, 0\leqslant y^*+\leqslant 2$

7. $-1\leqslant x^*, y^*\leqslant 1$

8. 2.5

9. The equation to be solved, $e^{-y}+4y=0$ must be solved by numerical analysis (trial and error)

10. $z=\pm 1$

11. See problem 4

12. See problem 5

13. $L(x, y, \lambda) = 6x + 8y + \lambda(8x^2 + 3y - 7)$

14.   There is an unbounded minimum. The necessary conditions yield c maximum at $x = -9/64$, $y = 3503/1536$.

15.   The second derivative with respect to both x and y is 2. Hence the original function is the sum of two convex functions and is convex.

16. $x = 3$, $y = 4$

17.   See Table 9.2

18. $L/2^{(n-3)}$

Chapter 10

1. $V = 5$

2a. $V = 10$

2b. $V = 8$

2c. $V = 8$

2d. $V = 5$

4. $x_1 = 20$, $x_2 = 20$

7.   4-5-6-1-3-2

8.   4-6-5-3-1-2

Index